D1531642

Centrifugation in Biology
and Medical Science

Centrifugation in Biology and Medical Science

PHILLIP SHEELER

*Department of Biology
California State University
Northridge, California*

A WILEY-INTERSCIENCE PUBLICATION

JOHN WILEY & SONS   New York · Chichester · Brisbane · Toronto

Library of Congress Cataloging in Publication Data:

Sheeler, Phillip.
  Centrifugation in biology and medical science.

  "A Wiley-Interscience publication."
  Includes index.
    1. Centrifugation. 2. Biological chemistry—
Technique. 3. Biology—Technique. I. Title.
[DNLM: 1. Centrifugation. QH324.9.04 S541c]

QP519.9.C44S53   574'.028   80-21744
ISBN 0-471-05234-5

Printed in the United States of America

10 9 8 7 6 5 4 3 2 1

To my wife, Annette

Preface

It would be difficult to overstate the value of centrifugation as a tool of biomedical research, for much of what we know today about the composition, properties, and functions of specific macromolecules, supermolecular complexes, subcellular organelles, and even whole cells is the consequence of studies carried out following centrifugal purification, isolation, and/or characterization of these materials. In this book, I attempt to present a comprehensive account of the diverse forms that centrifugation takes in biological and medical science, noting particular applications, advantages, limitations, and other characteristics of each approach. The book was written with the view toward its usefulness not only to researchers but also to advanced students whose interests lie in physiology, cellular and molecular biology, biochemistry, and related fields. It is intended that the book's thorough consideration of criteria that underlie successful centrifugal fractionations can serve as a guide to the design of experimental protocols that meet specific tasks or goals. Special attention is paid to recent developments in the field and the appendices bring together a wealth of data of general usefulness.

My interest in centrifugation began in 1965 when I had the good fortune to receive a postdoctoral fellowship to study at the Biology Division of the Oak Ridge National Laboratory under the guidance of Dr. Norman G. Anderson. At that time, Dr. Anderson's "Molecular Anatomy (MAN) Program" was devoted primarily to the development of zonal centrifuges and centrifugal fast analyzers. Although I had joined the laboratory to continue earlier studies on the mechanism by which iron is transferred across the plasma membrane of the developing red blood cell, I quickly became engrossed in the remarkable centrifugal developments being pioneered by Dr. Anderson and his associates. Fascinated by the seemingly unlimited applications of centrifugation in biomedical research and profoundly influenced by Dr. Anderson's inventive genius, my own research interests and activities have been directed since that time to centrifugal methodology.

Norman Anderson's manifold contributions have helped turn the modern centrifuge into one of the most powerful and versatile tools in the research laboratory, and it is the focal point of many analytical and preparative studies. The effectiveness with which centrifugation is used depends to a large extent on the researcher's awareness of the spectrum of techniques that are available and appreciation of the specific advantages and limitations of each approach. By establishing the fundamental principles of centrifugation and surveying the field generally, this book provides a foundation on which such expertise can be developed.

This book could not have been prepared without the help and contributions of a number of individuals and companies. First, thanks to Dr. Norman G. Anderson of the Argonne National Laboratory (U.S.A.) for his permission to reproduce a number of photographs of early zonal rotors. My appreciation is also extended to Dr. W. Howard Evans of the National Institute for Medical Research (United Kingdom) for his contributions of photos of the A-XII rotor. I am grateful also to Mac Lawrence (Beckman Instruments, Inc.), Ronald Ostrom, Keelin Fry, and Rodger Nelson (DuPont/Sorvall Instruments), Patrick W. Goulding and V. N. Musmanno (Pennwalt Corporation, Sharples-Stokes Division), Ellen P. Aquilina (Bausch and Lomb) and Brian Collins (Buchler Instruments Division of Searle Diagnostics, Inc.) for their generous contributions of photographs and pertinent information.

Much of the original artwork appearing in the book was prepared by my good friend and colleague Mark H. Doolittle. Over the years, my own work in the field of centrifugation has been supported by the manifold talents of Harry R. White, to whom I owe special thanks. Finally, I should like to express my appreciation to Mary M. Conway, Life Sciences Editor, Denise Hillhouse, Sr. Production Supervisor, and Carole Schwager, Editorial Supervisor for Wiley-Interscience publications for their help and encouragement during the writing and production of this book.

PHILLIP SHEELER

Northridge, California
December, 1980

Acknowledgment

Special thanks are extended to Mark H. Doolittle, who so masterfully prepared many of the illustrations in the book.

Contents

Centrifugation in Biology
and Medical Science

ONE

An Introduction

Centrifugation is one of the most important and widely applied research techniques in biochemistry, in cellular and molecular biology, and in medicine. In its various forms, centrifugation is regularly used to (1) remove cells or other suspended particles from their surrounding milieu on either a batch or a continuous-flow basis, (2) separate one cell type from another, (3) isolate viruses and macromolecules, including DNA, RNA, proteins, and lipids or establish physical parameters of these particles from their observed behavior during centrifugation, and (4) separate from dispersed tissue the various subcellular organelles including nuclei, mitochondria, chloroplasts, Golgi bodies, lysosomes, peroxisomes, glyoxysomes, plasma membranes, endoplasmic reticulum, polysomes, and ribosomal subunits. Indeed, much of our present knowledge concerning the morphology, chemical composition, and physiological functions of specific populations of cells in heterogeneous tissues and of the various subcellular organelles has been obtained either directly from analytical centrifugal techniques or by chemical or microscopic study following centrifugal isolation of these minute structures from disrupted tissues and cells.

It is important at the outset to clarify several terms regularly used either in connection with centrifugation as a technique or to classify different kinds of centrifuges; these are "ultracentrifuge," "analytical" ultracentrifuge, and "preparative" centrifuge (or ultracentrifuge). The term "ultracentrifuge" was introduced in 1924 by Theodor Svedberg to identify an instrument in which the sedimentation of amicroscopic colloidal particles could be directly observed and followed (Svedberg and Rinde, 1924). Svedberg's rationale was to make the identification of his centrifuge consonant with the recently developed "ultramicroscope"—a light microscope with special optics for rendering visible tiny particles not discernable by conventional light microscopy. It was not Svedberg's original

1

intent that the term "ultracentrifuge" suggest a centrifuge capable of producing very high rotational speeds; indeed, the rotors in Svedberg's 1924
ultracentrifuge reached only 10,000 rpm. However, in today's usage, the
term "ultracentrifuge" is generally reserved for centrifuges (analytical
or preparative) in which rotational speeds in excess of 25,000 rpm are
routine, with the rotor spinning in an evacuated chamber. (Centrifuges
operating at speeds below this are variously called "superspeed" or
"high-speed" centrifuges.)

Svedberg's ultracentrifuges were designed to characterize large molecules, especially proteins, on the basis of the behavior of these particles
in centrifugal fields. The centrifuges accommodated only small quantities
of sample and were equipped with an optical system that permitted the
progress of particle sedimentation to be viewed and photographed. An
analysis of their sedimentation behavior revealed certain physical properties of the particles even though individual particles were not actually
isolated or collected. Accordingly, the Svedberg instruments were known
as *analytical* ultracentrifuges in order to distinguish them from *preparative*
instruments used to collect sediments from particle suspensions.

The practical distinction between analytical and preparative centrifuges became less clear as new procedures evolved for using progressively
refined preparative instruments not only to quantitatively isolate particles,
but to yield analytical data from the behavior of the particles during the
isolation procedure. Even the optical system that for so many years
uniquely characterized the Svedberg-type analytical ultracentrifuge has
been modified and adapted to preparative instruments; moreover, special
transparent rotors (i.e., zonal rotors; see Chapter 6) can be used with
many preparative centrifuges, so that the investigator can directly observe
the course of particle separations.

For the most part, the term "analytical untracentrifuge" is now restricted to a particular series of instruments characterized by an elaborate
optical system and used either to establish physical parameters of particles
on the basis of precise measurements of sedimentation made during centrifugation or to accurately assess the purity (or heterogeneity) of biological and chemical samples prepared by other techniques. In effect,
these instruments are modern-day versions of the classical Svedberg concept. Other centrifuges are referred to as *preparative* instruments even
though they may be used in either analytical experiments or in the quantitative isolation of particles.

Over the years, a number of books have been devoted to the subject
of centrifugation, and these are listed at the end of this chapter. However,
most of these books have focused either on classical analytical ultracentrifugation, dealing at length with mathematical models for the behavior

of sedimenting particles, or on specific applications of a given preparative approach. The present work is of a much more general character and encompasses methods that are not considered in any depth in other books dealing with the general subject of centrifugation. It is intended that this book serve several roles for the reader:

1 To provide a practical mathematical foundation in the physical principles on which centrifugal methodology is founded.
2 To provide an awareness of the remarkable variety of task-specific centrifugal equipment and accessories available today and an appreciation of the origin and the development of the major forms of centrifugation being used today.
3 To draw attention to the diversity of experimental approaches that can be taken in order to attack both general and specific biomedical problems.

Using the information presented here, it is hoped that the reader will be in a better position to make more prudent decisions concerning what centrifugal methods should (or should not) be tried in an effort to attain one's experimental objectives.

Following the cited references that appear at the end of each chapter are lists of additional articles dealing with specific problems that may bear more directly on the reader's own experimental objectives.

EARLY HISTORY OF CENTRIFUGATION

The practical application of centrifugal force dates back more than a thousand years to the extraction of "tung oil," a substance used in paints and varnishes, from the seeds of *Aleurites cordata* trees. However, the first recorded biological study using a "centrifuge" is that of T. A. Knight in 1806, who showed that the roots and stems of seedlings oriented themselves to centrifugal force when grown at the circumferential edge of a rotating water wheel. By the 1870s, small hand-operated centrifuges capable of spinning two small cylindrical test tubes at speeds up to about 3000 rpm were being used by chemists for collecting precipitates, whereas larger but slower devices were used by dairyworkers for removing the sediment present in fresh milk and for cream separation. Carl Gustaf de Laval, a Swedish inventor whose main efforts were directed toward the design of steam-driven turbine engines, introduced the *continuous-action* centrifugal cream separator in 1878 and completely revolutionized the dairy industry. This marked the beginning of continuous-flow centrifugation and represented a major step forward in centrifuge technology. De Laval's basic approach to continuous centrifugal separations remains essentially unchanged even in modern instruments, and some of Svedberg's first analytical ultracentrifuges were modified versions of commercial de Laval machines. Although the electrical motor had been invented by Michael Faraday in 1822 and commercial electric motors were available by 1880, it was not until about 1910 that motor-driven centrifuges were built. Prior to 1910, centrifugal devices were either hand operated or driven by water or steam.

Small centrifuges were used in the 1880s by biologists to study properties of protoplasm and cellular inclusions. For example, the viscosity of protoplasm was investigated in 1880 by C. Dehnecke, who followed the movements of starch grains through the cytoplasm of plant cells subjected to small centrifugal forces. D. Mottier showed in 1899 that cellular inclusions visible with the light microscope formed strata within cells subjected to centrifugal force, suggesting that these inclusions possess different densities. Although several nineteenth century biologists effected the isolation of certain cellular inclusions, these isolations were achieved by noncentrifugal methods. Most notable, perhaps, was Friedrich Miescher's isolation of cell nuclei from leukocytes in 1868 and from sperm cells in 1874 by chemically digesting away the extranuclear cytoplasm.

In medicine, S. G. Hedin and J. Daland showed in the early 1890s that they could more rapidly determine the ratio of plasma volume to packed cell volume by subjecting whole blood to centrifugal force rather than allowing the cells to settle through the plasma for several hours under the influence of gravity. These early "hematocrit centrifuges" were small, hand-operated instruments geared to provide speeds up to 10,000 rpm and capable of forming the packed blood cell sediment in about 3 min. In 1902, A. E. Wright, a pioneer in blood cytology, introduced the use of specially prepared glass centrifuge tubes in which the leukocyte-rich "buffy coat" could be conveniently separated from the bulk of erythrocytes for subsequent microscopic study. Wright's work probably represents the first successful fractionation of whole cells using centrifugal force.

The focal point of biochemistry at the turn of the century was the investigation of the chemical and physical properties of protein solutions and other colloids. The discovery by Emil Fischer in 1902 that proteins consisted of chains of amino acids linked together through peptide bonds turned out to be a major impetus for further development of centrifugal technology. Fischer and most other notable chemists of that period believed that individual protein species (hemoglobin, albumin, etc.) were *polydisperse;* that is, a given protein molecule occurred in various sizes (i.e., polypeptide chain lengths). Reasoning that the polydispersity of colloidal suspensions might be amenable to study by measuring particle sedimentation rates, A. Dumanskii (Dumanskii, 1910; Dumanskii et al., 1913) attempted to correlate the results of ultramicroscopic studies of colloids with the behavior of the colloidal particles during centrifugation. Dumanskii's experiments, which were carried out with the use of a conventional centrifuge containing cylindrical tubes (and hence lacking "sector shape"), were unsuccessful because of the influence of convection on the sedimenting particles. The Svedberg had been concerned with colloid chemistry since 1905, and initially he shared the view that individual protein species were polydisperse. In 1923, Svedberg and J. B. Nichols built an "optical centrifuge"—the immediate forerunner of the analytical ultracentrifuge. The rotor of the optical centrifuge consisted of a metal tube encasing two cylindrical glass chambers. The tube was mounted directly on a small electric motor so that its long axis rotated in the horizontal plane. Narrow rectangular slots cut through the tube permitted the sedimentation of particles to be viewed and photographed during centrifugation. Rotor speed was limited to a few hundred revolutions per minute. Particle sedimentation rates could not be measured accurately in the optical centrifuge because of convective disturbances resulting from the lack of sector shape in the tube chambers and the absence of strict temperature control.

Later in 1923, Svedberg and H. Rinde built the first ultracentrifuge. The basic driving mechanism was constructed by using parts of a de Laval centrifugal cream separator. The cream separator bowl was replaced by the centrifuge rotor (constructed of brass) and the top of the separator modified to enclose the rotor in a hydrogen gas atmosphere that cooled the rotor during operation and reduced thermal convection in the samples subjected to analysis. The rotors used in this centrifuge were positioned on a vertical shaft, the lower portion of which formed a worm gear. The worm gear was rotated in the horizontal plane by a vertical gear wheel connected by a belt to an electric motor (Fig. 1-1). Svedberg soon replaced the original gear-driven ultracentrifuges by direct motor-driven machines.

To eliminate sedimentation artifacts resulting from convection currents within the sample compartments (called "cells"), the compartments were constructed with a sector shape; that is, the cell walls formed radii of the circle being swept out by the spinning rotor. The cells were fitted above and below with transparent windows permitting direct observation (and photography) during the course of sedimentation. Rotors attained speeds of 10,000 rpm (i.e., about 5000g) and were used to study the behavior of a number of colloids, including hemoglobin. Working with R. Fahraeus, Svedberg made the first estimations of the average molecular

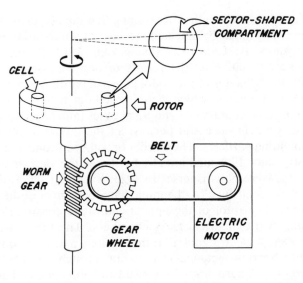

Fig. 1-1 Basic components of Theodor Svedberg's first analytical ultracentrifuge.

weight of hemoglobin (67,000) (Svedberg and Fahraeus, 1926). Because of the relatively low centrifugal forces produced in this centrifuge, the hemoglobin studies were necessarily *sedimentation equilibrium* experiments, in which particle sedimentation is balanced by diffusion, and therefore, could not provide information about molecular polydispersity or homogeneity. Such information could be provided only by *sedimentation velocity* methods, in which centrifugal force exceeds diffusion, so that a sedimenting boundary is formed. Sedimentation equilibrium and sedimentation velocity experiments are considered more fully later.

By 1926, Svedberg was also experimenting with centrifuges in which the rotor was driven by two small oil turbines positioned at either end of the rotor shaft. These rotors were spun about a horizontal axis at speeds up to 40,100 rpm (about 100,000g). In the same year, Svedberg received the Nobel prize in chemistry for his work on colloids. Using direct, motor-driven, and oil-turbine-driven ultracentrifuges, Svedberg continued his studies on colloids, and from a series of experiments conducted through 1931 drew two notable conclusions: (1) proteins may have molecular weights from thousands (e.g., egg albumin) to millions (e.g., snail hemocyanin); and (2) a given protein type (hemoglobin, albumin, hemocyanin, etc.) is not polydisperse but is homogeneous, with all molecules of that protein having the same molecular weight. The notion that proteins possessed well-defined molecular weights represented a departure from contemporary thinking and was received with considerable skepticism.

Between 1924 and 1940, Svedberg constructed and tested a number of rotors varying in size, shape, and the number and positions of their analytical cells. Some of these rotors were used at speeds in excess of 80,000 rpm. During the same period, E. Henriot and E. Huguenard pioneered the idea of using compressed air to spin centrifuge rotors. The air entered the centrifuge chamber from below and acted to support the rotor as well as to spin it; the rotors were small, solid devices that did not hold liquid samples for analysis (Henriot and Huguenard, 1925). By 1927, they reported attaining 660,000 rpm using the air-turbine approach (Henriot and Huguenard, 1927; Huguenard, 1927). Substituting compressed hydrogen gas for air and using a rotor of less than 1 cm in diameter, J. W. Beams and E. G. Pickels in 1935 achieved 1,560,000 rpm. The possibility of using the air-turbine approach for centrifuges and rotors suitable for biological and chemical studies was pursued in the late 1930s and early 1940s by E. G. Pickels and H. J. Bauer. Pickels and Bauer placed the air turbine in a separate compartment above the rotor chamber. A length of wire descended from the turbine into the rotor chamber and supported the rotor from above. Rotation of the turbine was translated through the wire to the rotor (Bauer and Pickels, 1936). The rotors used by Bauer and Pickels and the system for optical measurement of the sedimenting particles were similar to those developed by Svedberg.

Pickels, Bauer, J. L. Onceley, R. W. G. Wyckoff, and W. M. Stanley successfully employed the air-turbine analytical ultracentrifuge at speeds up to about 60,000 rpm in a series of studies on proteins and virus particles. In 1946, responding to the increasing interest in the analytical ultracentrifuge, Pickels formed the Special Instruments Company (SpInCo) for the commercial production of the instrument. Later this became the Spinco division of Beckman Instruments, Inc. For simplification and greater reliability, the air-turbine drive was replaced by an electric motor and gears, but in other respects the ultracentrifuge remained the same. Indeed, while the Spinco instrument (known as the "model E analytical ultracentrifuge") has undergone a number of modifications and improvements over the years, it still possesses much of the character of the original Pickels ultracentrifuge. Commercial analytical ultracentrifuges are also produced by other companies, including Measuring and Scientific Equipment, Ltd. (MSE) and Hereus-Christ, Inc.

The technological evolution of the analytical ultracentrifuge served as a catalyst for furthering the development of other kinds of centrifuges for chemical, biological, and medical study. The result is today's host of superspeed centrifuges, ultracentrifuges, and rotors for general- as well as special-purpose preparative, continuous-flow, and analytical applications. Many of the succeeding chapters of this book are devoted to a description of the development of these instruments and their practical applications.

Research and development in the field of centrifugation continues today at a rapid pace as the major centrifuge manufacturers strive to improve the performance of existing instruments and as independent researchers try new and often ingenious ways of attacking old problems. Novel and new drive mechanisms like the DuPont oil turbine and the gearless and beltless direct induction drive recently introduced by Beckman Instruments have pushed the operating speeds and centrifugal forces available to the general research laboratory to new highs (80,000 rpm and more than 600,000g). Recently, Katano and Shimizu (1979) described an experimental centrifuge that employs magnetic fields to both support and spin a small rotor; with their centrifuge, rotor speeds in excess of 12,-000,000 rpm (100,000,000g) have been obtained.

REFERENCES AND RELATED READING

Books Dealing Exclusively with Centrifugation

Birnie, G. D., and Rickwood, D., eds. *Centrifugal Separations in Molecular and Cell Biology*. Butterworths, London, 1978.

Bowen, T. J. *An Introduction to Ultracentrifugation*. Wiley-Interscience, London, 1970.

Hinton, R., and Dobrota, M. *Density Gradient Centrifugation*. North-Holland, Amsterdam, 1976.

McCall, J. S., and Potter, B. J. *Ultracentrifugation*. Bailliere-Tindall, London, 1973.

Rickwood, D., ed. *Centrifugation: A Practical Approach*. Information Retrieval, London, 1978.

Schachman, H. K. *Ultracentrifugation in Biochemistry*. Academic, New York, 1959.

Svedberg, T., and Pederson, K. O. *The Ultracentrifuge*. Oxford Clarendon Press, London, 1940.

Books with Sections Devoted to Centrifugation

Brewer, J. M., Pesce, A. J., and Ashworth, R. B. *Experimental Techniques in Biochemistry*. Prentice-Hall, Englewood Cliffs, N. J., 1974.

Cooper, T. G. *The Tools of Biochemistry*. Wiley-Interscience, New York, 1977.

Trautman, R. Ultracentrifugation. In *Instrumental Methods of Experimental Biology*, (D. W. Newman Ed.), MacMillian, New York, 1964.

Articles and Reviews

Aston, D. (1978) The evolution of liquid and gas centrifuges. *Endeavor,* **2,** 142.

Bauer, J. H., and Pickels, E. G. (1936) High speed vacuum centrifuge for filterable viruses. *J. Exp. Med.,* **64,** 503.

Beams, J. W. (1938) High speed centrifuging. *Rev. Mod. Phys.,* **10,** 245.

Beams, J. W. and Pickels, E. G. (1935) The production of high rotational speeds. *Rev. Sci. Instrum.,* **6,** 299.

Dumanskii, A. (1910) The influence of centrifugal force on the equilibrium of a chemical system. *J. Russ. Phys. Chem. Soc.,* **41,** 1306.

Dumanskii, A., Zabotinskii, E., and Vseyev, M. (1913) A method for determining the size of colloidal particles. *Z. Chem. Ind. Kolloide,* **12,** 6.

Gray, G. W. (June 1951), The ultracentrifuge. *Sci. Amer.*

Henriot, E., and Huguenard, E. (1925) High speed rotation. *C. R. Acad. Sci.,* **180,** 1389.

Henriot, E., and Huguenard, E. (1927) Large angular velocity derived from rotors without solid axle. *Journal de Physique et la Radium,* **8,** 443.

Huguenard, E. (1927) Great angular speeds obtained with rotors without solid axle. *Rev. Gen. des Sciences,* **38,** 565.

Katano, R., and Shimizu, S. (1979) Production of centrifugal fields greater than 100 million times gravity. *Rev. Sci. Instrum.,* **50,** 805.

Moore, D. H. Gradient centrifugation. In *Physical Techniques in Biological Research,* (2d ed.) Vol. II, Part B, D. H. Moore, Ed. Academic, New York, 1969.

Pederson, K. O. (1974) The analytical ultracentrifuge: Part I. Svedberg and the early experiments. *Fractions,* No. 1.

Svedberg, T., and Nichols, J. B. (1923) Determination of size and distribution of size of particle by centrifugal methods. *J. Amer. Chem. Soc.,* **45,** 2910.

Svedberg, T., and Rinde, H. (1924) The ultracentrifuge, a new instrument for the determination of size and distribution of particle in amicroscopic colloids. *J. Amer. Chem. Soc.,* **46,** 2677.

Svedberg, T., and Fahraeus, R. (1926) A new method for the determination of the molecular weight of the proteins. *J. Amer. Chem. Soc.,* **48,** 430.

Williams, J. W. (1974) The analytical ultracentrifuge: Part II. From the colloid experiments to DNA. *Fractions,* No. 2.

TWO

Some Fundamental Physical and Mathematical Concepts

The behavior of particles subjected to centrifugation depends on a number of interacting factors, including the conditions of centrifugation and the properties of the particles and their surroundings. In this chapter we consider some of the basic physical laws that influence sedimentation and derive a number of mathematical relationships that can be used to define experimental conditions and interpret or predict particle sedimentation behavior

GRAVITY

It was Galileo Galilei (1564–1642) who first systematically and scientifically investigated gravity as a natural phenomenon. Before Galileo (i.e., from the time of Aristotle), it was generally accepted that the velocity with which objects fall toward the earth is directly proportional to their weight, with heavy objects falling more rapidly than light objects. However, Galileo tested a number of prospective relationships between time t, distance fallen x, and velocity dx/dt and found that the relationship supported by experiment was

$$\frac{dx}{dt} = at \tag{2-1}$$

that is, the velocity attained by a falling object is proportional to the time interval during which the object falls. The proportionality constant a represents the rate of acceleration. Integration of equation 2-1 yields

$$x = \tfrac{1}{2} at^2 \tag{2-2}$$

Galileo recognized that the relationship derived between time and speed applied only under idealized conditions, *including the absence of air.*

Sir Isaac Newton (1642–1727, by strange coincidence born the year that Galileo died) furthered the study of falling objects, showing that their acceleration is produced by a *force,* namely, *gravity,* acting between the earth and the falling body. Acceleration due to gravitational force is constant for any given location on the earth's surface but varies from one location to another according to elevation and latitude. At sea level the gravitational acceleration constant is 977.99 cm/sec sec at 0° latitude, 980.60 cm/sec sec at 45° latitude, and 983.21 cm/sec sec at 90° latitude. The gravitational acceleration constant is customarily assigned the symbol g and for simplicity taken to be 980 cm/sec sec. Newton showed that the force acting on a body undergoing constant acceleration is equal to the product of its mass and its acceleration; that is,

$$F = ma \qquad\qquad (2\text{-}3)$$

(with F in dynes, m in grams, and a in cm/sec sec). For a body accelerated by the earth's gravity,

$$F = mg \qquad\qquad (2\text{-}4)$$

As an illustration of the relationships just described, let us determine the speed attained under idealized conditions by an object (initially at rest) falling from a height of 100 m. From equation 2-2, we obtain

$$x = \tfrac{1}{2} at^2 = \tfrac{1}{2} gt^2 \qquad\qquad (2\text{-}5)$$

(with x in centimeters, t in seconds, and g = 980 cm/sec sec). Therefore,

$$10,000 = \frac{980}{2}\,(t^2)$$

and t = 4.52 sec (i.e., it takes 4.52 sec for the object to fall to earth). Using equation 2-1,

$$\frac{dx}{dt} = at = gt = (980)(4.52) = 4430$$

That is, at the time the object reaches the surface of the earth, it is falling at 4430 cm/sec. In making these calculations, we have ignored the resistance that would be offered by air. The effect of this resistance is to progressively diminish acceleration such that during a prolonged fall, the falling object approaches a *limiting velocity.* For example, a parachutist in free fall (i.e., parachute not yet opened) reaches a limiting velocity of about 90 m/sec [about 200 miles per hour (mph)] within a few minutes.

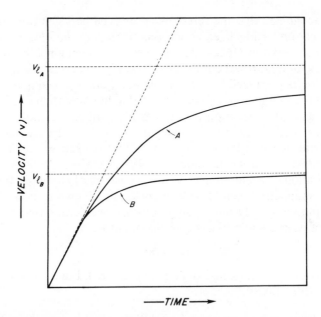

Fig. 2-1 Relationship between time and velocity for objects falling (or sedimenting) through a medium offering resistance. A heavy object (*A*) approaches a limiting velocity over a longer time interval than a light object (*B*).

In the absence of air resistance (e.g., in a vacuum), This velocity would be reached in just 9 sec. Moreover, after the parachute is opened (thereby further increasing air resistance), a new and *lower* limiting velocity will be approached. For objects that have the same general shape, the limiting velocity is higher for the heavier object and is approached over a longer time interval, as depicted in Fig. 2-1.

In the same manner that air resists the acceleration of an object by gravity, the liquid that surrounds particles in suspension resists their sedimentation during centrifugation; consequently, sedimenting particles also approach a limiting velocity.

CENTRIFUGAL FORCE

Consider an object *P* of mass *m* moving in a horizontal circular path at a constant speed (i.e., at a specific number of revolutions per unit of time) and at a fixed distance *x* from the axis of rotation (Fig. 2-2). Although the speed is constant, the direction of movement is continuously changing. Since *velocity* is defined in terms of the direction of motion, and this direction is changing, the velocity is continuously changing and the object is said to manifest constant acceleration. To restrict the movement to a circular path, a force must be exerted in a horizontal plane toward the axis of rotation. In the absence of such a force, the object would move off along a tangent to the circle (dashed arrow in Fig. 2-2). The force, called *centrifugal force,* is given by the same relationship as equation 2-3, where a is the acceleration toward the axis of rotation.

The speed with which the object rotates around the axis may be expressed in radians per second ω (where one complete revolution sweeps out 2π radians). Referring to Fig. 2-2, with respect to any starting point on the circle (e.g., its intersection with the Y axis), the angle swept out by P in time t would be ωt. The X and Y coordinates of the particle are related to ω (called the *angular velocity*) and radius of rotation x in the

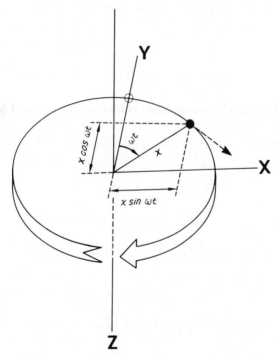

Fig. 2-2 The changing x, y coordinates of a particle moving at angular velocity ω in a circular path of radius x may be used to show that the particle's acceleration equals $\omega^2 x$. See text for derivation.

following manner:

$$X = x \sin(\omega t) \tag{2-6}$$
$$Y = x \cos(\omega t) \tag{2-7}$$

At any instant the velocity in the X direction is dX/dt and in the Y direction dY/dt; thus

$$\frac{dX}{dt} = \omega x \cos(\omega t) \tag{2-8}$$

$$\frac{dY}{dt} = -\omega x \sin(\omega t). \tag{2-9}$$

Since the velocity of the particle is continuously changing, the resulting acceleration may be expressed in terms of its X and Y components. Accordingly,

$$\frac{d(dX/dt)}{dt} = \frac{d^2X}{dt^2} = -\omega^2 x \sin(\omega t) \tag{2-10}$$

$$\frac{d(dY/dt)}{dt} = \frac{d^2Y}{dt^2} = -\omega^2 x \cos(\omega t). \tag{2-11}$$

Note that the X and Y components of the acceleration are negative, indicating that they act in the direction *opposite* to that in which X and Y are measured—namely, toward the center of the circle. The magnitude of the acceleration toward the center is the resultant of these two components. That is,

$$a = \sqrt{\left(\frac{d^2X}{dt^2}\right)^2 + \left(\frac{d^2Y}{dt^2}\right)^2} \tag{2-12}$$

$$= \sqrt{\omega^4 x^2 \sin^2(\omega t) + \omega^4 x^2 \cos^2(\omega t)} \tag{2-13}$$

$$= \sqrt{\omega^4 x^2 [\sin^2(\omega t) + \cos^2(\omega t)]} \tag{2-14}$$

Since $\sin^2(\omega t) + \cos^2(\omega t) = 1$,

$$a = \sqrt{\omega^4 x^2} = \omega^2 x \tag{2-15}$$

Thus the force accelerating particle P of mass m toward the axis of rotation is

$$F = ma = m\omega^2 x \tag{2-16}$$

During centrifugation, the spinning rotor and its parts are held at a fixed radial distance by centrifugal force, but suspended particles, which are not under such constraints, are free to move away from the axis of rotation. The force on the sedimenting particles is also given by equation 2-16.

It is customary in connection with centrifugation to refer to the force that causes a particle to move radially *away* from the axis of rotation as centrifugal force, whereas a force directed *toward* the axis is called *centripetal* force. In the same connection, movement of particles away from the axis of rotation is called *centrifugal movement,* and movement toward the axis is called *centripetal movement.*

RELATIVE CENTRIFUGAL FORCE—*g* FORCE

Usually, the value cited for the force applied to a suspension of particles during centrifugation is a relative one; that is to say, it is compared with the force that the earth's gravity would have on the same particles. It is called *relative centrifugal force* (RCF), where

$$\text{RCF} = \frac{F_{\text{centrifugation}}}{F_{\text{gravity}}} = \frac{m\omega^2 x}{mg} = \frac{\omega^2 x}{g} \qquad (2\text{-}17)$$

Using the value $g = 980$ cm/sec sec,

$$\frac{\omega^2}{g} = \frac{[(2\pi)(\text{rpm})/60]^2}{980} \qquad (2\text{-}18)$$

$$= 1.119 \times 10^{-5} (\text{rpm})^2 \qquad (2\text{-}19)$$

Therefore,

$$\text{RCF} = 1.119 \times 10^{-5} (\text{rpm})^2\, x \qquad (2\text{-}20)$$

To determine the RCF in effect during centrifugation, it is necessary to measure the speed (in revolutions per minute) and the distance between the axis of rotation and the particles. Since RCF is a ratio of two forces, it has no units; however, it is customary to follow the numerical value of the RCF with the symbol *g*. This indicates that the RCF being applied is a specific multiple of the force exerted by the earth's gravity.

Example What is the relative centrifugal force at 10 cm from the axis of rotation in a rotor spinning at 20,000 rpm?

$$\text{RCF} = 1.119 \times 10^{-5} (20,000)^2 (10)$$

$$= 4.476 \times 10^4 \quad \text{or} \quad 44,760g$$

It is to be noted that the RCF applied during centrifugation, readily calculated by using equation 2-20, is independent of the particles being sedimented.

Coriolis Force   In addition to centrifugal force, particles in suspension (and the body of suspending fluid itself) within a spinning rotor are subjected to *Coriolis* force. The Coriolis force, which results from the inertia of the liquid and the suspended particles, is a small force directed at right angles to both the axis of rotation and the direction of the centrifugal force. For a rotor rotating clockwise, the force acts to deflect particles in a counterclockwise direction (and vice versa). Under nearly all experimental conditions, the Coriolis force is very small in comparison with the centrifugal force and may be ignored. Since the Coriolis force is an inertial force, its effects are magnified when the rotor's speed changes (e.g., during acceleration and deceleration). For centrifugation involving density gradients (see Chapters 4 and 5), Coriolis forces can create problems if the rotor is accelerated and/or decelerated too rapidly. Coriolis forces also become more significant when the rotor's contents are displaced centrifugally or centripetally *while the rotor is spinning*—for example, when carrying out density gradient separations in zonal rotors or performing continuous-flow centrifugation. These special cases are considered later in the book.

PARTICLE SEDIMENTATION RATE

While centrifugal force acts to accelerate a particle away from the axis of rotation, the radially sedimenting particle is also subjected to additional forces including frictional force, the force of buoyancy, and gravitational force. The directional relationships between these forces are summarized in Fig. 2-3.

Frictional Force   The movement of a particle through a liquid (or through a gas, such as air) is influenced by the liquid's *viscosity*. The viscosity of a liquid may be determined by measuring the time required for a given volume of the liquid to flow through a capillary tube under the influence of gravity. The viscosity is calculated by using *Poiseuille's equation,* namely,

$$\eta = \frac{\pi h g \rho_M r_c^4 t}{8LV} \qquad (2\text{-}21)$$

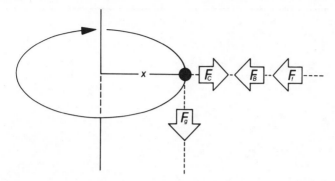

Fig. 2-3 Directions of forces acting on a particle moving in a horizontal circular path through a medium offering resistance: F_C, centrifugal force; F_B, buoyant force; F_f, frictional force; F_g, force due to gravity.

where h = average height of liquid pressure head
       g = gravitational acceleration constant
       ρ_M = density of liquid (i.e., of *medium*)
       r_c = radius of capillary tube
       t = time of flow (in seconds)
       L = length of capillary tube
       V = volume of liquid flowing through the tube.

The units of viscosity η are *poise* (grams per centimeter per second) or *centipoise* (grams per meter per second). See Chapter 4, Figs. 4-17 and 4-20 for the relationships between the concentration, the density, and the viscosity of several solutions used in centrifugation. It is to be noted that the viscosity of a solution also varies with its temperature (Fig. 2-4).

In 1856, Sir Gabriel Stokes (1819–1903) published the results of his extensive studies on the frictional resistance encountered by a particle moving through a viscous medium. For rigid, nonhydrated, spherical particles of radius r migrating through a liquid of viscosity η at velocity dx/dt, the force due to frictional resistance is given by

$$F_f = 6\pi\eta r \, (dx/dt) \tag{2-22}$$

This relationship is known as *Stokes' law* and applies under conditions in which the particles are large compared to the molecules that comprise the liquid medium and are not present at a concentration so high as to influence the liquid's viscosity.

Most biological particles are not spherical, necessitating modification of equation 2-22 in estimating the frictional force. However, a great many biological particles have shapes that may be approximated by *ellipsoids*

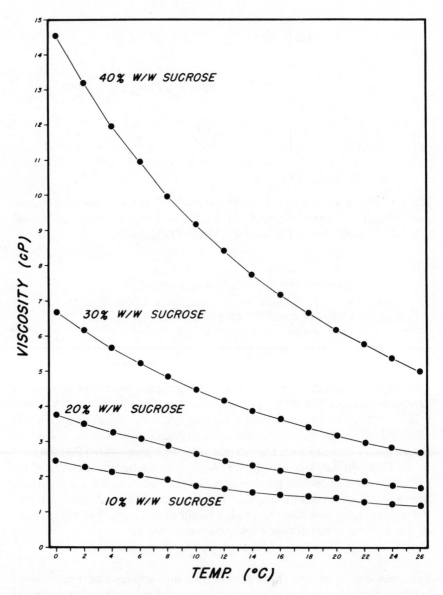

Fig. 2-4 Relationship between temperature and viscosity of sucrose solutions of various concentrations.

of revolution, that is, by the spheroids formed by rotating an ellipse around its major or minor axes (see Fig. 2-5). Rotation around the major axis produces a *prolate* ellipsoid, whereas rotation about the minor axis produces an *oblate* ellipsoid. The frictional resistance experienced by an ellipsoid is greater than that for a sphere of the same volume and increases with increasing *axial ratio* or *eccentricity* (i.e., *A:B* in Fig. 2-5). The frictional ratios for various prolate and oblate ellipsoids are given in Table 2-1, where f/f_0 is the ratio of the frictional resistance encountered by an ellipsoid (i.e., f) and by a sphere (i.e., f_0) of the same volume.

Since frictional resistance varies according to particle volume as well as shape, it should be noted that prolate and oblate ellipsoids that have the same axial ratio as well as equal major and minor axes experience different frictional resistances because *both their frictional ratios and sizes differ.* For example, prolate and oblate ellipsoids that have an axial ratio of 4:1 and major and minor axes of 4 and 1 units, respectively, will differ in volume by a factor of 4 (the oblate ellipsoid being larger; see Fig. 2-5). Table 2-2 compares in arbitrary units of length the major and minor axes of equal-volume prolate and oblate ellipsoids and spheres; for such particles, frictional resistance will vary only according to frictional ratios.

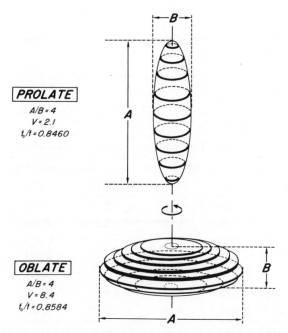

PROLATE
A/B = 4
V = 2.1
f_0/f = 0.8460

OBLATE
A/B = 4
V = 8.4
f_0/f = 0.8584

Fig. 2-5 Representative prolate and oblate ellipsoids of revolution.

Table 2-1 Frictional Ratios for Ellipsoids of Revolution

Axial Ratio (A/B)	Frictional Ratio (f/f_0)	
	Prolate	Oblate
1 (Sphere)	1.000	1.000
2	1.044	1.042
3	1.112	1.105
4	1.182	1.165
5	1.255	1.224
6	1.314	1.277
8	1.433	1.374
10	1.543	1.458
15	1.784	1.636
20	1.996	1.782
50	2.946	2.375
100	4.067	2.974
1000	13.16	6.369

Table 2-2 Major and Minor Axes of Various Ellipsoids that Have the Same Volume as a Sphere of Unit Radius

Axial Ratio (A/B)	Prolate Ellipsoid		Oblate Ellipsoid		Sphere Radius (r)
	Major Axis (A)	Minor Axis (B)	Major Axis (A)	Minor Axis (B)	
1	2.000	2.000	2.000	2.000	1.000
2	3.174	1.587	2.520	1.260	1.000
3	4.160	1.387	2.885	0.962	1.000
5	5.848	1.170	3.420	0.684	1.000
10	9.283	0.928	4.308	0.431	1.000
100	43.08	0.431	9.283	0.093	1.000
1000	200.0	0.200	20.00	0.020	1.000

sphere volume $= \frac{4}{3}\pi r^3$; prolate ellipsoid volume $= \frac{4}{3}\pi (A/2)(B/2)^2$; oblate ellipsoid volume $= \frac{4}{3}\pi (A/2)^2 (B/2)$.

Stokes' equation for frictional force (equation 2-22) may be modified to account for ellipsoidal particles; in other words,

$$F_f = 6\pi\eta r \; dx/dt \; \frac{f}{f_0} \qquad (2\text{-}23)$$

where r is the radius of the sphere that has the same volume as the ellipsoid. For particles that have shapes other than that of a sphere or an ellipsoid, equation 2-23 requires further modification. For example, the frictional force encountered by sheetlike structures would be appreciably greater.

Force of Buoyancy   The Greek mathematician Archimedes (287–212 B.C.) showed that the weight of an object suspended in a liquid is effectively diminished by the weight of the liquid being displaced by the object (this is known as *Archimedes' principle*). The object is said to be acted on by a *buoyant force,* which may be defined as follows:

$$F_B = V_P \, (\rho_M) \, g \qquad (2\text{-}24)$$

where V_P is the volume of the object (or particle) and ρ_M is the density of the displaced liquid. Note that the force of buoyancy is *independent of the shape* of the object. Since the volume of a particle is equal to its mass m divided by its density ρ_P, equation 2-24 may also be written

$$F_B = \frac{m}{\rho_P} \, (\rho_M) \, g \qquad (2\text{-}25)$$

Equations 2-24 and 2-25 apply to particles suspended in liquids at rest (i.e., subjected only to gravity). However, if the particle suspension is rotated in a centrifuge, the buoyant force on the particles is magnified in proportion to the square of the angular velocity and the distance between the suspended particles and the axis of rotation; that is,

$$F_B = \frac{m}{\rho_P} \, (\rho_M) \, \omega^2 x \qquad (2\text{-}26)$$

Effects of Gravity on Particles During Centrifugation   Even at low rotor speeds, the centrifugal force exerted on suspended particles is considerably greater than the force due to gravity. For example, even at a speed of only 1000 rpm, the centrifugal force on particles 12 cm from the axis of rotation is more than 134 times that due to gravity. Therefore, during centrifugation in a rotor spinning about an axis perpendicular to the earth's surface (the usual case), the movement of the suspended particles parallel to the axis (i.e., vertically downward) is minor in comparison with their movement away from the axis. However, if the centrifugal movement of particles is terminated by their reaching and accumulating at the margins of the spinning rotor, subsequent downward movement by gravity may be notable. The latter notion is pursued later in the book.

For simplicity and convenience, we can ignore the effects of the earth's gravitational force on movements of suspended particles during centrifugation and concentrate instead on the consummate effects of centrifugal force (F_C), frictional force (F_f) and the bouyant force (F_B).

Measuring dx/dt   Consider a hypothetical situation in which a single spheroidal particle is suspended in a body of liquid (e.g., water) that instantaneously undergoes the transition from rest to rapid rotation about an axis. When the liquid is at rest, the only forces acting on the particle are the downward force due to gravity and the upwardly directed buoyant and frictional forces, all of which are very small (see above). Now, the centrifugal force produced by transition to the rotating state causes the particle to move away from the axis of rotation. As the velocity of its radial movement (represented by dx/dt) increases (from its initial value of zero), so do the frictional and buoyant forces opposing this movement. As a result, the particle quickly (almost instantaneously) reaches a *limiting velocity*, as the centrifugal force is balanced by the forces of friction and buoyancy. That is, $d(dx/dt)/dt$ approaches zero, and dx/dt becomes constant when

$$F_C = F_f + F_B \qquad (2\text{-}27)$$

The value of dx/dt may be determined by substituting the values of F_C, F_f, and F_B given in equations 2-16, 2-23, and 2-26 into equation 2-27. Accordingly,

$$m\omega^2 x = 6\pi\eta r \frac{dx}{dt}\frac{f}{f_0} + \frac{m}{\rho_P}(\rho_M)\,\omega^2 x \qquad (2\text{-}28)$$

Substituting the volume of a sphere (i.e., $\frac{4}{3}\pi r^3$) times its density for its mass in equation 2-28, we obtain

$$(\tfrac{4}{3}\pi r^3)(\rho_p)(\omega^2 x) = \left[6\pi\eta r \frac{dx}{dt}\frac{f}{f_0}\right] + (\tfrac{4}{3}\pi r^3)(\rho_M)\,\omega^2 x \qquad (2\text{-}29)$$

By factoring and transposing, we obtain

$$\tfrac{4}{3}\pi r^3\,(\rho_P - \rho_M)\,\omega^2 x = 6\pi\eta r \frac{dx}{dt}\frac{f}{f_0} \qquad (2\text{-}30)$$

Solving for dx/dt, we obtain

$$\frac{dx}{dt} = \frac{\tfrac{4}{3}\pi r^3\,(\rho_P - \rho_M)\,\omega^2 x}{6\pi\eta r\,(f/f_0)} \qquad (2\text{-}31)$$

which simplifies to

$$\frac{dx}{dt} = \frac{2r^2\,(\rho_P - \rho_M)}{9\eta\,(f/f_0)}\,\omega^2 x \tag{2-32}$$

THE SEDIMENTATION COEFFICIENT

For a spheroidal particle of given size and density suspended in a liquid of known density and viscosity, the terms r, ρ_P, ρ_M, η, and f/f_0 will be constant, so that dx/dt becomes directly proportional to $\omega^2 x$ (Fig. 2-6). A convenient expression called the *sedimentation coefficient s* is defined by the relationship

$$s = \frac{dx/dt}{\omega^2 x} \tag{2-33}$$

Therefore, for a spheroidal particle

$$s = \frac{2r^2\,(\rho_P - \rho_M)}{9\eta\,(f/f_0)} \tag{2-34}$$

The sedimentation coefficient of a particle can be calculated if experimental measurements are substituted into an integrated form of equa-

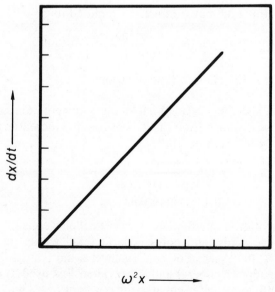

Fig. 2-6 Linear relationship between dx/dt and $\omega^2 x$.

tion 2-33. If we let x_1 be the distance in cm between the axis of rotation and a particle at time t_1 and let x_2 be the distance at time t_2, then by transposition of terms in equation 2-33, that is,

$$s \, dt = \frac{1}{\omega^2} \frac{dx}{x} \tag{2-35}$$

and integrating between the limits set above, that is,

$$s \int_{t_1}^{t_2} dt = \frac{1}{\omega^2} \int_{x_1}^{x_2} \frac{dx}{x} \tag{2-36}$$

we obtain

$$s \, (t_2 - t_1) = \frac{1}{\omega^2} (\ln x_2 - \ln x_1) = \frac{1}{\omega^2} \left(\ln \frac{x_2}{x_1} \right) \tag{2-37}$$

and

$$s = \frac{\ln(x_2/x_1)}{\omega^2(t_2 - t_1)} = 2.303 \frac{\log_{10}(x_2/x_1)}{\omega^2(t_2 - t_1)} \tag{2-38}$$

Since

$$\omega = \frac{2\pi \, (\text{rpm})}{60} = 0.105 \, (\text{rpm}) \tag{2-39}$$

$$\omega^2 = 10.97 \times 10^{-3} \, (\text{rpm})^2 \tag{2-40}$$

and thus equation 2-38 may be further simplified to

$$s = \frac{2.1 \times 10^2 \log_{10} (x_2/x_1)}{(\text{rpm})^2 \, (t_2 - t_1)} \tag{2-41}$$

where $(t_2 - t_1)$ is the elapsed time in seconds.

Example (see also Fig. 2-7)   What is the sedimentation coefficient of a particle whose distance from the axis of rotation increases from 6.2 cm to 11.7 cm at 54,000 rpm in 2 hr?
    From equation 2-41, we have

$$s = \frac{2.1 \times 10^2 \log_{10} (11.7/6.2)}{(54,000)^2 \, (2)(60)(60)} = 2.76 \times 10^{-12} \text{sec}$$

The sedimentation coefficients of many cellular constituents such as proteins, nucleic acids, and polysaccharides fall within the range 1×10^{-13} to 200×10^{-13} sec. For convenience, a unit called the *Svedberg unit* (after Theodor Svedberg) and abbreviated S is used to denote sedimentation coefficients and is equal to the constant 10^{-13} *sec*. Hence most

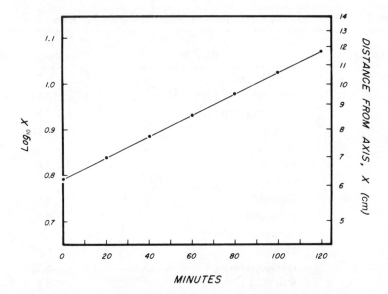

Fig. 2-7 Distances of a particle from the axis of rotation at various times of centrifugation (see example problem).

cellular proteins have sedimentation coefficients between 1 and 200S. In the above example, the value of s could be given as 27.6S.

It is possible to show that the sedimentation coefficient of a particle in seconds is also a reasonable approximation to the amount of time that it would take the particle to approach its limiting velocity. In other words, a particle having a sedimentation coefficient of 100S and an initial velocity of zero would accelerate for about 10^{-11} sec before approaching its limiting velocity.

It is customary to express sedimentation coefficients in terms of particle behavior at 20°C in pure water (i.e., $s_{20,w}$) either by performing analyses under these conditions or by appropriately adjusting the calculated s value when other conditions are used. The correction may be made using the following relationship:

$$s_{20,w} = (s_{T,m}) \frac{\eta_{T,m} (\rho_P - \rho_{20,w})}{\eta_{20,w} (\rho_P - \rho_{T,m})} \tag{2-42}$$

where $s_{T,m}$ is the experimentally determined sedimentation coefficient using medium m and temperature T, $\eta_{T,m}$ is the viscosity of the medium at T, $\eta_{20,w}$ is the viscosity of water at 20°C, ρ_P is the density of the particle, $\rho_{T,m}$ is the density of the medium at T, and $\rho_{20,w}$ is the density of water at 20°C. The $s_{20,w}$ values for a variety of particles are given in Table 2-3.

Table 2-3   Sedimentation Coefficients of Some Biological Particles

Group	Particle	$s_{20,w}$ (Svedberg Units)
Proteins	Cytochrome c	1.7
	Myoglobin	1.82
	Egg-white lysozyme	1.9
	Insulin	1.95
	Trypsin	2.5
	Pepsin	2.8
	G-Actin	3.7
	Collagen	4.0
	Hemoglobin	4.1
	Albumin (plasma)	4.5
	Myosin	6.4
	Hemerythrin	6.75
	Fibrinogen	7.6
	Hemocyanin (octopus)	58.7
	Hemocyanin (snail)	100
Nucleic acids	Transfer-RNA	4
	Histone messenger RNA	9
	T_7 Bacteriophage DNA	30
	Prokaryotic ribosomal RNA	5
		16
		23
	Eukaryotic ribosomal RNA	5
		5.8
		18
		28
Viruses	Turnip yellow mosaic virus	106
	Poliomyelitis virus	154
	Tobacco mosaic virus (TMV)	180
	Rabbit papilloma virus	280
	Simian virus 40 (SV40)	240
	T_7 Bacteriophage	490
	T_2 Bacteriophage	700
	Influenza virus	700

Table 2-3 (continued)

Group	Particle	$s_{20,w}$ (Svedberg Units)
Cellular inclusions	Apoferritin	17.6
	Ferritin	up to 63
	Glycogen	up to 10^5
	Starch grains	10^6 to 10^7
Cellular organelles	Nucleosomes	11
	Prokaryote ribosome subunits	30 (small)
		50 (large)
	Eukaryote ribosome subunits	40 (small)
		60 (large)
	Prokaryote ribosomes (monomers)	70
	Eukaryote ribosomes (monomers)	80
	Polysomes	
	Dimers	123
	Trimers	154
	Tetramers	183
	Pentamers	211
	Hexamers	237
	Membrane fragments	10^2 to 10^4
	Plasma membranes	up to 10^5
	Smooth endoplasmic reticulum	10^3
	Rough endoplasmic reticulum	10^3
	Lysosomes	4×10^3 to 2×10^4
	Peroxisomes	4×10^3
	Mitochondria	1×10^4 to 7×10^4
	Chloroplasts	10^5 to 10^6
	Nuclei	10^6 to 10^7
	Whole cells	10^7 to 10^8

Close inspection of equation 2-32 reveals that the parameters determining sedimentation rate (i.e., dx/dt) fall into three major categories: (1) the physical conditions of centrifugation (i.e., the values of ω and x), (2) the nature of the suspending medium (i.e., its density and viscosity), and (3) the properties of the suspending particles (their sizes and densities). The relationship between dx/dt and $\omega^2 x$ has already been discussed (see above). Increases in either (or both) the density or the viscosity of the suspending medium are accompanied by a decrease in dx/dt, but the relative effects of these two factors depend on the specific density–viscosity relationship of the suspending liquid (see Fig. 4-20, in Chapter 4). A dramatic illustration of this is given in Fig. 2-8, which compares the sedimentation rates of ideal spherical particles of density 1.6 in sucrose and CsCl solutions at 20°C. In sucrose, particle sedimentation rates drop sharply in increasingly dense solutions principally because of the rapid increase in the viscosity of the medium. This is seen in Fig. 2-8 by comparing the sucrose curve with the dashed line, which depicts the change in dx/dt in a hypothetical solution in which the viscosity remains constant at increasing densities. The CsCl curve more closely approximates the

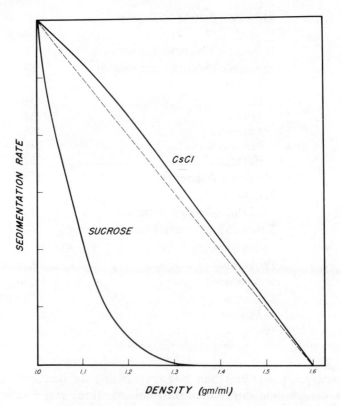

Fig. 2-8 Comparison of the relative sedimentation rates for a spherical particle that has a density of 1.6 in sucrose and CsCl solutions of different density. The two curves are different because the viscosities of sucrose and CsCl solutions vary differently in relation to solute concentration. The dashed curve shows the sedimentation rates that would occur if the viscosities of sucrose and CsCl solutions did not change.

latter situation since there is little change in CsCl viscosity with increasing density in the range 1.0 to 1.6. In fact, in the early portion of the CsCl curve, the values of dx/dt diminish less rapidly than in our hypothetical solution because the viscosity is actually reduced at increasing CsCl concentrations. We return to these relationships later in the book in connection with discussions of *density gradient centrifugation*.

The particular properties of spheroidal particles that affect dx/dt are size (i.e., radius) and density. Of these two parameters, sedimentation rate is more dependent on size since dx/dt varies in proportion to the *square* of the particle's radius but is a *first-order* function of the particle's density. These interrelationships are also pursued more fully later.

DIFFUSION

Diffusion is the bulk movement of solute molecules or particles from a region of higher concentration to a region of lower concentration. Such movement occurs as a consequence of random molecular motion or Brownian movement. Since the sedimentation of suspended particles is in effect a means of concentrating the particles, it is opposed by diffusion.

The rate at which diffusion takes place is given by *Fick's law* (A. E. Fick, 1829–1901), which states

$$\frac{dP}{dt} = - DA \frac{d(P)}{dx} \qquad (2\text{-}43)$$

where dP/dt is the amount of substance P diffusing through the cross-sectional area A in time t, $d(P)/dx$ is the concentration gradient, and D is the *diffusion coefficient* and takes on a specific value for each species of molecule or type of particle. In general, an inverse relationship exists between the size of a molecule or particle and its diffusion coefficient, so that small molecules or particles undergo more rapid diffusion than do larger ones. The effects of diffusion in opposing the sedimentation of particles during centrifugation are greater in the case of molecules such as proteins than for larger particles such as the subcellular organelles. For example, at about 8,000 rpm, the centrifugal movement of a protein of 60,000 molecular weight is sufficiently slow as to be counterbalanced by the centripetal diffusion of the protein. Consequently, the protein remains uniformly distributed through the suspending medium. In contrast, when particles like mitochondria, lysosomes, and peroxisomes, are subjected to centrifugation, diffusion can essentially be ignored. The effects of diffusion are a special concern when conducting sedimentation equilibrium or sedimentation velocity experiments using an analytical ultracentrifuge. Further discussion of the effects of diffusion is deferred to Chapter 8.

REFERENCES AND RELATED READING

Books

McCall, J. S., and Potter, B. J. *Ultracentrifugation*. Bailliere-Tindall, London, 1973.

Schachman, H. K. *Ultracentrifugation in Biochemistry*. Academic, New York, 1959.

Trautman, R. Ultracentrifugation. In *Instrumental Methods of Experimental Biology*, D. W. Newman, Ed. MacMillan, New York, 1964.

Articles and Reviews

Bruner, R., and Vinograd, J. The evaluation of standard sedimentation coefficients of sodium RNA and sodium DNA from velocity data in concentrated NaCl and CsCl solutions. *Biochim. Biophys. Acta,* **108,** 18 (1965).

Kaempfer, R., and Meselson, M. Sedimentation velocity analysis in accelerating gradients. In *Methods in Enzymology,* Vol. 20, S. P. Colowick and N. O. Kaplan, Eds. Academic, New York, 1971.

THREE

Centrifugal Fractionation of Tissues and Cells

BASIC APPROACHES AND INSTRUMENTATION

Prior to 1930 few, if any, efforts were directed toward the use of centrifugation for *isolating* subcellular organelles and particles. Rather, in a popular approach of the time, whole cells were subjected to centrifugal force and the microscopically observed redistribution of the organelles and particles within the cells used as the basis for making inferences about the density, viscosity, or other physical properties of the cytoplasm and the organelles themselves. What few attempts were made to isolate subcellular organelles from disrupted tissue were carried out for the most part by chemical *extraction* procedures using a variety of exotic combinations of solutes and solvents.

The pioneering analytical ultracentrifugal studies of soluble proteins and other macromolecules carried out in the 1920s have already been noted (see Chapter 1), but it was not until the 1930s and the innovative studies of Bensley and Hoerr (1934), Hill (1937), Behrens (1938 and 1939), and Granick (1938) that centrifugation was used in a systematic way to isolate subcellular structures and to obtain particle-free cytoplasm. By the late 1940s, principally as a result of the model approaches carefully worked out by Nobel laureate Albert Claude (1946a and 1946b) and his students G. H. Hogeboom, W. C. Schneider, and G. E. Palade (Hogeboom et al., 1948), centrifugation became the focal point of tissue fractionation procedures.

DIFFERENTIAL CENTRIFUGATION

The basic approach originally used by Behrens, Bensley, Claude, and others and used today in variously modified and improved forms is termed *differential centrifugation* or *differential pelleting*. The procedure involves the stepwise removal of different classes of particles in an heterogeneous mixture by successive centrifugations at increasing RCF. Typically, the tissue to be fractionated is first disrupted to free the subcellular particles, and the suspension (usually called a *homogenate*) is subjected to low-speed centrifugation to sediment the largest (or densest) particles present. Following this, the unsedimented material (called the *supernatant*) is carefully separated from the sediment (called the *pellet*) by decantation and is subjected to higher-speed centrifugation, thus sedimenting particles of somewhat smaller size (or lower density) and forming a new pellet and new supernatant. The sequence is repeated several times, and each centrifugation is carried out at successively higher speeds (and usually for longer periods of time) until a whole series of pellets (and a final supernatant) are obtained that may then be employed in further experimentation and analysis. Figure 3-1 depicts differential centrifugation

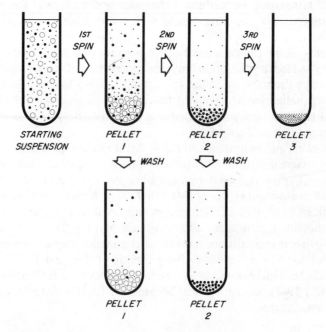

Fig. 3-1 Successive stages of *differential* centrifugation.

in its simplest form. Often, each of the pellets is resuspended in fresh medium and recentrifuged (i.e., the particles are "washed") one or more times in an effort to obtain a higher degree of particle purity (see below).

BASIC CENTRIFUGES AND ROTORS

Superspeed and Ultraspeed Centrifuges As noted in Chapter 1, two different types of preparative centrifuges may be distinguished: *superspeed centrifuges* and ultraspeed centrifuges or *ultracentrifuges*. Superspeed centrifuges generally operate at speeds to about 20,000 rpm, usually do not require evacuation of the rotor chamber, and either drive the rotor directly or through belts or gears. Ultracentrifuges can be operated at much greater speeds (up to 65,000 or 75,000 rpm); thus the rotor chamber must be evacuated of air to reduce friction and permit accurate rotor temperature control. In most ultracentrifuges the rotor is driven either by a motor and a set of gears or by an oil- or air-turbine system; very recently Beckman Instruments introduced a gearless induction motor drive. Table 3-1 lists some of the more popular research-quality superspeed and ultraspeed preparative centrifuges, and two of these are shown in Fig. 3-2.

Swinging-Bucket Rotors The most popular and widely used superspeed and ultraspeed centrifuge rotors are the *swinging-bucket* and *fixed-angle* rotors. The swinging-bucket (or *horizontal*) rotor (Fig. 3-3) consists of a central *harness* to which three, four, or six *buckets* are anchored through pivots. Each bucket accepts a *centrifuge tube* containing the sample. For swinging-bucket rotors used in superspeed centrifuges, the harness and buckets are generally enclosed in a *wind shield* to reduce air friction during spinning. The buckets used in ultracentrifuges are sealed with a screw cap and a rubber gasket so that the reduced air pressure resulting from evacuation of the rotor chamber during centrifugation does not cause evaporation (i.e., boiling) of the sample. During acceleration of the rotor, each bucket reorients from the vertical to the horizontal position and particles in the suspension sediment in the direction of the tube's long axis (Fig. 3-4).

Rotor and Tube Materials Early rotors such as the Svedberg rotors were made of steel and occasionally brass. The high density of these materials and the resulting high rotor weight produces an appreciable load on the centrifuge drive and significantly limits operating speeds. Most commercial rotors are now made partly or entirely of aluminum or titanium. As

Table 3-1   Common Superspeed and Ultraspeed Centrifuges

	Model	Maximum Operating Speed (rpm)
Superspeed centrifuges		
Beckman Instruments	J-21[a]	21,000
(Spinco Division)	J-21B[a]	21,000
	J-21C[a]	21,000
	J2-21	21,000
Dupont/Sorvall	RC-2[a]	20,000
Instruments	RC2-B[a]	20,000
	RC-5[a]	20,000
	RC-5B	20,000
International Equipment Co. (IEC)	B-20A	20,000
Ultraspeed centrifuges		
Beckman Instruments	L5-50B	50,000
(Spinco Division)	L5-65B	65,000
	L5-75B	75,000
	L8-55	55,000
	L8-70	70,000
	L8-80	80,000
DuPont/Sorvall	OTD-50 (OTD-50B)	50,000
Instruments	OTD-65 (OTD-65B)	65,000
	OTD-75 (OTD-75B)	75,000
International Equipment Co. (IEC)	B-60	60,000

[a] These are no longer manufactured, but many are still in routine use.

can be seen from Table 3-2, which compares the mechanical properties of various materials used in the construction of rotors, *titanium* is a superior choice because of its high strength:density ratio. Certain special applications rotors contain parts molded or machined from various plastics.

Centrifuge tubes were previously made of either glass or stainless steel, but these materials have effectively been replaced by different plas-

Table 3-2 Some Mechanical Properties of Materials Used in the Construction of Centrifuge Rotors

Material	Density (gm/cc)	Ultimate Strength (gm/cm^2)	Strength: Density Ratio
Plastic	1.41	1107	785
Aluminum	2.79	2159	774
Titanium	4.84	6088	1258
Steel	7.99	7915	991

Fig. 3-2 Popular ultraspeed and superspeed centrifuges. *Above*, OTD-75B ultracentrifuge, 75,000 rpm (DuPont/Sorvall Instruments); *Next page*, J2-21 superspeed centrifuge, 21,000 rpm (Beckman Instruments). (Photographs by permission of E. I. DuPont and Co. and Beckman Instruments, Inc.)

Fig. 3-2   continued

tics, including nylon, polyethylene, polypropylene, polycarbonate, polyallomer, cellulose nitrate, and cellulose acetate. Among these, polycarbonate is especially popular because of its strength and glasslike
transparency. Most plastic centrifuge tubes must be completely filled (or
nearly so) to avoid tube collapse at high centrifugal forces.

Fixed-Angle Rotors   *Fixed-angle* (or *angle head*) rotors are generally simpler in design than are swinging-bucket rotors (Fig. 3-3). In this type of
rotor, the centrifuge tubes are held at a specific and constant (i.e.,
"fixed") angle relative to the horizontal plane; that is, the tubes do not
reorient between the vertical and horizontal positions. Generally, the
angle of incline is between 15 and 30°. The "buckets" consist of 4 to 12
(or more) cylindrical openings machined into a solid, contoured metal
block. When fixed-angle rotors are used in the evacuated chamber of an
ultracentrifuge, either the individual tubes are sealed or the entire array
of buckets is sealed off by a cover and a set of o-rings or gaskets. A list
of some of the more popular swinging-bucket and fixed-angle rotors used
in superspeed and ultraspeed centrifuges may be found in the Appendixes
at the end of the book.

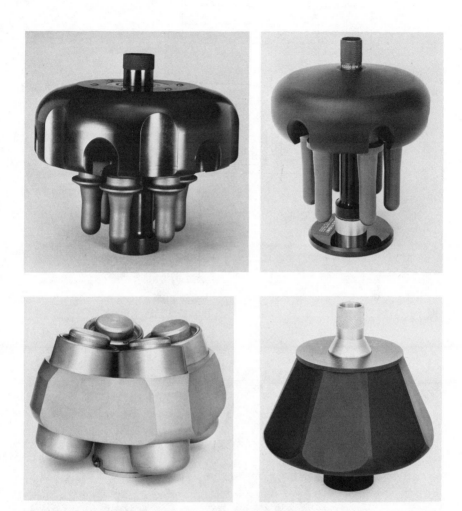

Fig. 3-3 Swinging-bucket (top) and fixed-angle (bottom) rotors. *Top left*, AH 650 (50,000 rpm); *top right*, SW 40Ti (40,000 rpm); *bottom left*, type 15 (15,000 rpm); bottom right, type 75Ti (75,000 rpm). (Top left photograph courtesy of E. I. DuPont and Company; others courtesy of Beckman Instruments, Inc.)

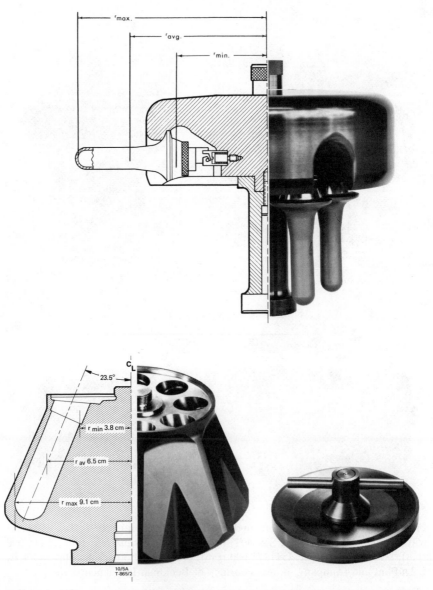

Fig. 3-4 Cross sections through a swinging-bucket (model AH627) and fixed-angle (model T865) rotor. In swinging bucket rotors, particles travel in the direction of the tube's long axis after the tube has swung into the horizontal position. In fixed-angle rotors, the particles sediment against the outer sloping wall of the tube and then down the wall to the bottom. See also Fig. 3-5. (Photographs courtesy of E. I. DuPont and Company.)

38

CONVECTION IN SWINGING-BUCKET AND FIXED-ANGLE ROTORS

Convection occurs whenever uniform suspensions of particles are sedimented in conventional swinging-bucket and fixed-angle rotors. The term "convection" means the *bulk* movement of solute and/or solvent within the centrifuge tube. The most common cause of *unwanted* convection is *temperature variation* in different parts of the centrifuge tube. However, when swinging-bucket and fixed-angle rotors are used to pellet particles, convection resulting from local *density inversions* materially increases the rate of pellet formation. The tubes used in swinging-bucket rotors have parallel walls, but particle sedimentation occurs radially during centrifugation. Consequently, the concentration of particles (and the density of the suspension) increases at the tube walls, whereas the particle concentration (and hence the suspension density) decreases in the center of the tube (Fig. 3-5A). The resulting convection takes the form of bulk

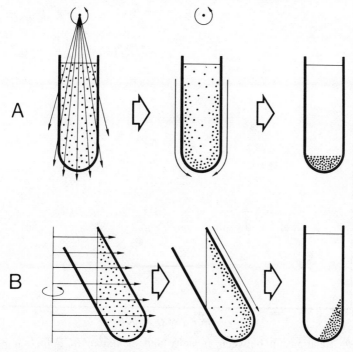

Fig. 3-5 Convection in swinging-bucket (*A*) and fixed-angle (*B*) rotors leads to bulk movement of particles along the walls of the centrifuge tube. See text for details.

movement of particles along the tube walls toward the bottom of the centrifuge tube.

Convection is even more pronounced in fixed-angle rotors where the centrifuge tubes are never oriented in the horizontal plane. Particles that sediment radially away from the axis of rotation travel only a short distance (approximately equal to the diameter of the tube) before encountering and accumulating at the tube wall (Fig. 3-5B). The resulting density increase along one edge of the tube is followed by the rapid movement of the particles toward the bottom of the tube by convection.

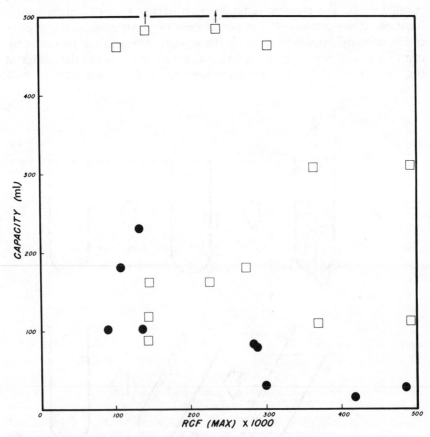

Fig. 3-6 Relationship between capacity and maximum RCF for some representative swinging-bucket (closed circles) and fixed-angle (open squares) rotors. In general, fixed-angle rotors have higher capacities and can be operated at greater speeds than can swinging-bucket rotors.

In general, the more pronounced convective forces at play in fixed-angle rotors make these rotors more efficient than swinging-bucket rotors for pelleting suspended particles. (This, of course, presumes that other factors, such as rotor speed and dimensions, are the same.) Fixed-angle rotors generally accommodate more sample than do swinging-bucket rotors (i.e., the number of tube positions and their capacities is greater), and since they have no moving parts and are machined and contoured from a single block of metal, fixed-angle rotors can be operated at higher speeds (Fig. 3-6). Consequently, the fixed-angle rotor is generally the instrument of choice for rapid and quantitative pelleting. On the other hand, swinging-bucket rotors may be preferred when dealing with suspensions of particles that may stick to the tube walls or may be damaged by sliding down the wall. In addition, swinging-bucket rotors are preferred when particles are to be separated in *density gradients* (see Chapters 4 and 5).

Not all suspended particles are subjected to the same RCF in swinging-bucket or fixed-angle rotors since the radial position of a particle varies between minimum and maximum values (Fig. 3-4). When expressing the RCF used during centrifugation at a given speed with a specific rotor, it is customary to note the maximum RCF, or g_{max}, although g_{ave} (i.e., the RCF at r_{ave}) or g_{min} (i.e., the RCF at r_{min}) may also be cited. For example, in the T-865 fixed-angle rotor (see Fig. 3-4) operating at 65,000 rpm, the RCF values would be $179,655g_{min}$, $307,300g_{ave}$, and $430,230g_{max}$. Because of the angle of incline, r_{min} usually is a larger value in fixed-angle rotors than in swinging-bucket rotors capable of the same maximum speed. This, too, tends to make fixed-angle rotors more efficient for pelleting suspended particles.

NATURE OF TISSUE FRACTIONS PRODUCED BY DIFFERENTIAL CENTRIFUGATION

The approaches to tissue fractionation taken by Behrens, Bensley, and others in the early 1930s had as a major aim the isolation from disrupted cells of one or more identifiable components that could then be physically and chemically characterized. The principal method of analysis was light microscopy. In this sense, their approach was strictly preparative and is to be contrasted with the more analytical procedures of Claude and his successors, Schneider, Hogeboom, and Palade, who were concerned with the purity of centrifugally isolated fractions, examining their enzymatic activities and questioning whether certain properties were unique to specific types of subcellular particles or were more broadly distributed.

Fundamental to any assessment of the properties of a given cellular entity and the assignment to that component of a specific biological activity is the need to be certain that one is dealing with "pure" fractions. In this direction, Claude repeatedly washed each of the pellets produced by his differential approach in an effort to remove any contaminating particles. Each washing was either added to the previous supernatant or separately analyzed and a complete "balance sheet" maintained for the distribution of activities in all collected fractions as well as the original, unfractionated sample.

THE FOUR MAJOR DIFFERENTIAL FRACTIONS

In the method introduced by Claude, Hogeboom, Schneider, and Palade, the following major differential fractions are produced when subjecting a tissue homogenate (e.g., liver tissue) to successive centrifugations at increasing RCF:

1 *The Nuclear Fraction*   The pellet produced by centrifugation of the original homogenate at 600g for 10 minutes.
2 *The Mitochondrial Fraction*   The pellet obtained when the post-nuclear supernatant (i.e., the supernatant of the first centrifugation) is centrifuged at 5000g for 10 minutes.
3 *The Microsomal Fraction*   The pellet produced when the postmitochondrial supernatant is centrifuged at 50,000g for 60 minutes.
4 *The Cytosol or Soluble Phase*   The supernatant of the last centrifugation.

The names originally assigned to the four major fractions were based primarily on microscopic examination of the material. For example, the "nuclear fraction" was so named because the major particulate elements observed were clearly the nuclei of the tissue cells. It is to be emphasized that the term "microsomal fraction" used to describe the final pellet was chosen because the particles present were extremely small (i.e., "micro-" = small and "-some" = body) and were generally unidentifiable. There is, therefore, no specific subcellular organelle called a *microsome;* rather, the expression is strictly an operational term used in connection with the centrifugal procedure. Claude, Hogeboom, Schneider, and Palade recognized that cell components other than those for which a fraction was named were present in that fraction; that is, the fractions were not pure. For example, the nuclear fraction clearly contained undisrupted whole cells and blood cells in addition to cell nuclei. Nonetheless, over the years, large numbers of researchers chose to identify their collected differential fractions as "nuclei," "mitochondria," and "microsomes," erroneously inferring that these were the only structures present. Table 3-3 lists some of the major cell structures now recognized to routinely be included in the four major differential fractions.

Markers As noted above, initially the most important analytical tool used to identify the components of a given centrifugal fraction was the light microscope. With improving technology, this gave way to the use of the electron microscope, which provides much greater detail and indeed renders identifiable many constituents of the microsomal fraction. However, an even more important tool is the use of "markers"—molecules, especially enzymes, known to be concentrated in one cell fraction. Indeed, the emphasis on the use of markers rather than morphological examination as the most reliable indicator of the types of particle present in the fractions produced by a given centrifugal scheme led to the discovery of new subcellular organelles, notably the lysosomes and peroxisomes. The importance of the prudent selection of markers and the need for an accurate, tabulated assessment of their distribution in the fractions and subfractions prepared by centrifugation cannot be overemphasized, and this imperative has been driven home repeatedly by Nobel laureate Christian de Duve, the most esteemed champion of this approach (see,

Table 3-3 Major Cell Fractions Produced During Differential Centrifugation of a Tissue Homogenate

Name of Fraction	Content
Nuclear fraction	Nuclei
	Plasma membranes
	Whole (undisrupted) tissue cells
	Blood cells
Mitochondrial fraction	Mitochondria
	Lysosomes
	Peroxisomes
Microsomal fraction ("microsomes")	Endoplasmic reticulum
	Golgi bodies
	Ribosomes and polysomes
Soluble phase ("cytosol")	Soluble enzymes
	Lipid
	Small molecules

for example, de Duve, 1964 and de Duve, 1971). Table 3-4 lists some of the more useful and reliable markers used in the identification of the components present in the tissue and cell fractions produced during centrifugation.

Nuclear Fraction   The nuclear fraction contains two major components: cell nuclei and pieces of the plasma membrane. Depending on the tissue used, the extent of organ perfusion, and the vigor with which the tissue homogenate is prepared, whole cells and blood cells may also be present. Numbers of mitochondria and other smaller cellular components also contaminate the nuclear fraction, perhaps through transient binding to the nuclei (see also below).

Mitochondrial Fraction   Mitochondria constitute the bulk of the mitochondrial fraction (originally called the *large granule fraction* by Claude), but also present are quantities of lysosomes and peroxisomes, and their amounts depend on the type of tissue being fractionated and the amount of shear force used to initially disrupt the tissue (e.g., lysosomes are particularly fragile and are disrupted if the tissue homogenization procedure is harsh).

Microsomal Fraction   The microsomal fraction is a mixture of small, particulate components of the cell. The endoplasmic reticulum is the major source of the microsomes yielding "rough" (i.e., ribosome-laden) and "smooth" (i.e., ribosome-free) vesicles through a pinching-off process that accompanies tissue homogenization. The membranes of the cell's Golgi bodies also are recovered in this fraction and so too are the ribosomes and the polyribosomes. Some fragments of the plasma membrane are recovered in the microsomes as well as pieces of the outer mitochondrial membrane; the latter are apparently stripped from some of the mitochondria during tissue homogenization. Small mitochondria, lysosomes, and peroxisomes may also contaminate this fraction. In liver tissue, glycogen particles are a major microsomal component.

Cytosol or Soluble Phase   The cytosol or soluble phase consists primarily of nonparticulate, dissolved substances, principally enzymes, salts, and water, less that quantity lost by adsorption to particles pelleted at earlier stages of differential centrifugation.

An increase in the purity of the pellets produced at each stage of differential centrifugation can be obtained by successively washing and recentrifuging the sedimented material; however, as should be noted in Fig. 3-1, any small, slowly sedimenting particles that initially reside near the bottom of the centrifuge tube will inevitably be recovered with the

Table 3-4 Enzymes and Other Substances Used as Markers During Subcellular Fractionation

Component	Marker	EC Number
Nuclei	DNA	
	DNA polymerase	2.7.7.7
	RNA polymerase	2.7.7.6
	NAD pyrophosphorylase	2.7.7.1
Plasma membranes	5'-Nucleotidase	3.1.3.5
	Leucine aminopeptidase	3.4.1.1
	Aminopeptidase	3.4.1.2
	Alkaline phosphodiesterase	3.1.4.1
	Mg^{++}-stimulated ATP′ase	3.6.1.4
	Leucyl-naphthylamidase	3.4.11.1
	Nucleotide triphosphatase	3.6.1.3
	Adenylate cyclase	4.6.1.1
	Cholesterol	
Mitochondria	Succinate dehydrogenase	1.3.99.1
	Cytochrome oxidase	1.9.3.1
	Glutamate dehydrogenase	1.4.1.2
	Monoamine oxidase	1.4.3.4
	Cytochrome-c oxidoreductase	1.6.99.1
Lysosomes	Acid phosphatase	3.1.3.2
	Aryl sulfatase c	3.1.6.1
	Phosphodiesterase-II	3.1.4.1
	β-glucuronidase	3.2.1.31
	Acridine orange	
Peroxisomes	Catalase	1.11.1.6
	Uric acid oxidase	1.7.3.3
	Peroxidase	1.11.1.7
Glyoxysomes	Glycollate oxidase	1.1.3.1
	Glyoxylate reductase	1.1.1.26
Golgi bodies	UDP galactose: N-acetyl glucosamine galactosyl transferase	2.4.1.38
	Sialyl transferase	2.4.99.1
Chloroplasts	Chlorophyll	
	Ribulose diphosphate carboxylase	4.1.1.39
Endoplasmic reticulum	Glucose 6-phosphatase	3.1.3.9
	NADPH-cytochrome c reductase	1.6.2.3
Cytosol	Aldolase	4.1.2.7
	Phosphoglucomutase	2.7.5.1
	Hexokinase	2.7.1.1

larger, rapidly sedimenting particles that form the bulk of the pellet during centrifugation. In effect, absolute purity cannot be achieved by differential centrifugation, and this deficiency created the need for and set the foundation for the development of density gradient techniques to be considered in the following chapters.

REFERENCES AND RELATED READING

Books
Behrens, M., Zell- und Gewebetrennung. In *Handbuch der Biologischen Arbeitsmethoden*, Vol. 5, (by E. Abderhalden, Ed.) Urban and Schwarzenburg, Berlin, 1938, Part 10, No. 2, p. 1363.

de Duve, C., and Grant, J. K., Eds. *Methods of Separation of Subcellular Structural Components*. Cambridge University Press, Cambridge, England, 1963.

de Duve, C., General principles (of enzyme cytology). In *Enzyme Cytology*, D. B. Roodyn, Ed. Academic, London, 1967.

Evans, W. H. *Preparation and Characterization of Mammalian Plasma Membranes*. North-Holland, Amsterdam, 1979.

Hinton, R., and Dobrota, M. *Density Gradient Centrifugation*. North-Holland, Amsterdam, 1976.

Trautman, R. Ultracentrifugation. In *Instrumental Methods of Experimental Biology*, D. W. Newman, Ed. MacMillan, New York, 1964.

Articles and Reviews
Behrens, M. (1939) Uber die Verteilung der Lipase und Arginase zwischen Zellkern und Protoplasma der Leber. *Z. Physiol. Chem.*, **258**, 27.

Bensley, R. R. (1943) The chemical structure of cytoplasm. *Biol. Symp.*, **10**, 323.

Bensley, R. R., and Hoerr, N. L. (1934) Studies on cell structure by the freezing–drying method. VI. The preparation and properties of mitochondria. *Anat. Rec.*, **60**, 449.

Claude, A. (1946a) Fractionation of mammalian liver cells by differential centrifugation. I. Problems, methods and preparation of extract. *J. Exp. Med.*, **84**, 51.

Claude, A. (1946b) Fractionation of mammalian liver cells by differential centrifugation. II. Experimental procedures and results. *J. Exp. Med.*, **84**, 61.

de Duve, C. (1964) Principles of tissue fractionation. *J. Theoret. Biol.*, **6**, 33.

de Duve, C. (1971) Tissue fractionation, past and present. *J. Cell Biol.*, **50**, 20.

Granick, S. (1938) Quantitative isolation of chloroplasts from higher plants. *Amer. J. Bot.* **25**, 558.

Hill, R. (1937) Oxygen evolved by isolated chloroplasts. *Nature*, **139**, 881.

Hogeboom, G. H., Schneider, W. C., and Palade, G. E. (1948) Cytochemical studies of mammalian tissues. *J. Biol. Chem.*, **172**, 619.

FOUR

Density Gradient Centrifugation

PREPARING GRADIENTS AND MEASURING DENSITY

As noted in Chapter 3, *differential centrifugation* in which each successive pellet is resuspended and recentrifuged (i.e., washed) several times provides one means for separating different families of particles present in a mixture. This approach is especially effective if the families of particles differ greatly in sedimentation coefficient. However, for families of particles having similar sedimentation coefficients, the pellets produced by differential centrifugation will be contaminated; that is, mixed in with the particles comprising the bulk of any pellet will be quantities of particles of lower sedimentation coefficients (see Fig. 3-1). The resolving power of centrifugation when used to fractionate particle mixtures took a major step forward with the introduction of the *density gradient* approach.

BASIC CONCEPT

The basic idea behind the density gradient approach is depicted in Fig. 4-1. The mixture of particles to be separated is carefully layered onto the surface of a vertical column of liquid the density of which progressively increases from top to bottom. Although the particles in the suspension are individually denser than the liquid at the top of the gradient, the *average* density for the sample (i.e., particles plus suspending liquid) is lower; only under such conditions could the sample zone be supported by the top of the density gradient. During centrifugation, the particles

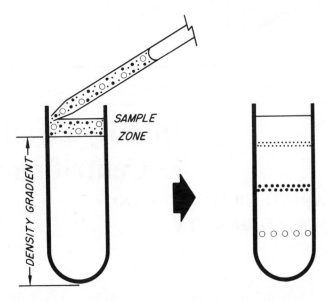

Fig. 4-1 Basic thesis of density gradient centrifugation. Sample containing a mixture of particles is layered onto the surface of the density gradient (left), and after centrifugation (right) the particles are distributed through the gradient as a series of zones.

sediment through the gradient toward the bottom of the centrifuge tube, and their sedimentation rates at any instant are determined by equation 2-32. In effect, at least during short runs in "shallow" density gradients (see below), the particles form zones or bands in the tube in order of their sedimentation coefficients. Following centrifugation, the separated zones are collected along with the gradient as a series of fractions either through the bottom of the tube (e.g., by puncturing the tube) or by upward displacement through a cap inserted into the top of the tube. The various procedures used to collect the gradient and entrained particle zones are discussed more fully later in the chapter.

ORIGINS OF DENSITY GRADIENT CENTRIFUGATION

The use of density gradients has become almost routine in centrifugal fractionations of particle mixtures, yet the historical development of this approach is somewhat difficult to trace. Credit for the introduction of density gradient centrifugation as a means of separating and isolating cells and subcellular components has been laid claim to or variously credited to M. Behrens (in 1938), to E. G. Pickels (in 1943), to N. G. Anderson (in 1949), to H. J. Kahler and B. J. Lloyd (in 1951), and to M. K. Brakke (in 1951). Some confusion stems from the fact that in the past the phrase "density gradient centrifugation" has been used to describe a number of approaches that are fundamentally quite different. In the late 1930s Behrens (1938) determined the densities of particles present in dessicated, powdered tissue extracts by suspending the particles in benzene and layering them onto density gradients prepared in centrifuge tubes by mixing benzene and carbon tetrachloride in continuously changing proportions. During centrifugation at 4000 rpm for up to 20 min, the particles sedimented through the organic liquid density gradients to positions where the particles and gradient had the same specific gravity (or density). Using current terminology, Behrens' work constitutes true density gradient centrifugation (in particular, a form called *isopycnic* density gradient centrifugation, discussed in Chapter 5). Behrens is appropriately credited with the first use of "continuous" density gradients to fractionate tissues.

Because they basically are simpler in design, fixed-angle rotors capable of attaining speeds high enough to sediment very small particles and macromolecules were developed several years in advance of their swinging-bucket counterparts. In 1943 Pickels sought to measure particle sedimentation rates and assess the homogeneity of particle suspensions using a fixed-angle rotor; the rationale was to follow the change in particle concentration at various levels of the centrifuge tube in the course of centrifugation. In an effort to reduce the anomolies that occur during sedimentation in fixed-angle rotors (see Chapter 3), Pickels suspended the particles within a sucrose density gradient. The density gradient served only to stabilize the column of particles against convection, and during centrifugation the particles did not form zones or bands (indeed, this was not the goal). Pickels' approach would not be considered density gradient centrifugation in the modern sense.

In unpublished studies in 1949 Anderson attempted the centrifugal fractionation of liver tissue homogenates by layering the samples onto density gradients formed using glycogen. Whereas Anderson's work probably represents the first use of *aqueous* density gradients for centrifugal fractionations akin to the type pioneered by Behrens, Brakke, whose studies with viruses were published in 1951 (see below), is generally accorded such credit.

In 1951 Kahler and Lloyd described a swinging-bucket rotor capable of very high speeds and studied its effectiveness in evaluating the sedimentation rates and homogeneity of latex particles. Like Pickels, Kahler and Lloyd suspended their particles in density gradients (in this instance, gradients formed by using glycerol) to reduce convection anomolies. They found that particle sedimentation rates in the horizontal tubes of the swinging-bucket rotor more closely approximate mathematically predicted values than do rates obtained by using a fixed-angle rotor. Neither rotor yielded credible results if the latex particles were not suspended within a density gradient. It was in the same year that Brakke successfully fractionated potato yellow-dwarf virus particles by layering virus concentrates over sucrose density gradients in tubes and centrifuging for several hours in a low-speed swinging-bucket rotor. Brakke's density gradients were prepared by successively layering sucrose solutions of different density in the centrifuge tube and then allowing several days of diffusion to smooth out the gradient before applying the sample to its surface. Brakke is generally credited as the first person to perform density gradient centrifugation in a manner closely approximating that used today.

PREPARATION OF DENSITY GRADIENTS

Density gradients are used not only for particle separations in the centrifuge tubes of conventional rotors, but also in special-purpose instruments such as *zonal* rotors, *continuous-flow* rotors, and *sta-put* devices; the latter density gradient procedures are considered in subsequent chapters of the book. The fundamentals of density gradient preparation are the same in nearly all applications, but for the time being the discussion will center around gradient generation in tubes.

Density gradients may be subdivided into two main categories according to the means of preparation: *step* gradients and *continuous* gradients.

STEP GRADIENTS

Step gradients are prepared by successively layering solutions of different density in the centrifuge tube and then layering the sample to be fractionated on top of the last "step." This is depicted in Fig. 4-2, which also shows the volume–density profile that exists in the tube at the completion of the loading operation. Frequently, an *overlay* is deposited above the sample to eliminate the variable thickness of the sample zone caused by formation of a meniscus. During centrifugation, the abrupt density changes across the interfaces between steps are somewhat smoothed out by diffusion (broken curve in Fig. 4-2). In a number of instances, application of the sample to the top of the step gradient is delayed for several

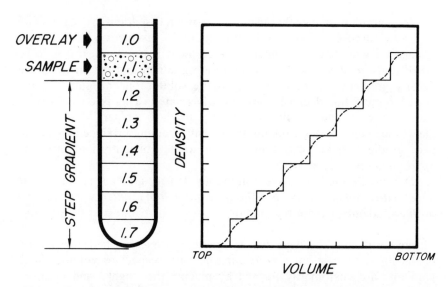

Fig. 4-2 Step gradients. The numbers in the centrifuge tube represent the density values of each step in the gradient, the density of the sample, and the density of the overlay. Before and/or during centrifugation, the steps are smoothed by diffusion.

hours or several days to allow diffusion to smooth out the gradient profile.

Special density gradient-generating equipment is not required for production of step gradients. The gradient may be built up in the centrifuge tubes by carefully layering one step on another, beginning with the densest step, by use of a pipette. Alternatively, the gradient can be formed beginning with the least dense step by depositing each layer at the bottom of the centrifuge tube through a narrow canula.

CONTINUOUS GRADIENTS

Continuous density gradients are gradients in which the density changes smoothly and continuously from one limit or extreme to another. Such gradients can be produced from step gradients by allowing sufficient time for diffusion to smooth out the steps, but continuous gradients are normally prepared directly by using special devices known as *gradient makers* or *gradient engines*.

The limiting density solutions selected for use with a gradient maker also represent solute *concentration* limits, and therefore the gradient produced is a *concentration gradient* as well as a density gradient. For example, a gradient maker that uses 10% w/w sucrose (density = 1.04) and 50% w/w sucrose (density = 1.23) as its limits produces a gradient that ranges in some specific manner between these two extremes. It is to be noted that the shape of the concentration gradient and the shape of the density gradient are identical only if the solute concentration and the corresponding liquid density are linearly related. This is only approximately true for most solutes regularly used to prepare density gradients (see below). Accordingly, the density gradient produced by a "linear" gradient maker varies linearly in solute concentration, but only approximately linearly in density.

The most commonly used continuous density gradient is one that is *linear with respect to volume,* although *exponential gradients* and *variable profile* gradients are not uncommon.

Linear Gradients   A linear density gradient is one in which the density of the liquid changes at a *constant* rate with respect to volume. Such gradients are customarily formed by placing the "light" and "dense" gradient limits in two interconnected chambers of the same shape and capacity and allowing the contents of one chamber to flow into the other (and be mixed there) at the rate dV/dt (where V = volume and t = time), while the mixture is withdrawn from the second chamber at the rate $2(dV/dt)$. Such a system in its simplest form is shown in Fig. 4-3.

The concentration of the stream of gradient emerging from the mixing chamber is given by

$$C_t = C_M + (C_R - C_M)\frac{V_t}{2V_i} \tag{4-1}$$

where C_t = concentration of gradient being delivered at any time t
C_R = concentration of reservoir liquid
C_M = starting concentration of mixing chamber
V_t = volume of gradient already withdrawn at time t
V_i = original volume of each chamber

Figure 4-4 shows a simple, "open-type" device for producing linear gradients; in this device the contents of the mixing chamber are stirred from above by using a rapidly vibrating paddle.

The following examples illustrate the use of equation (4-1):

Example 1   Suppose that the mixing chamber contains 30 ml of 0% w/w sucrose and the reservoir contains 30 ml of 40% w/w sucrose (i.e., a linearly increasing density gradient is being prepared). Then at that point in time when a volume of 30 ml has been delivered, the concentration

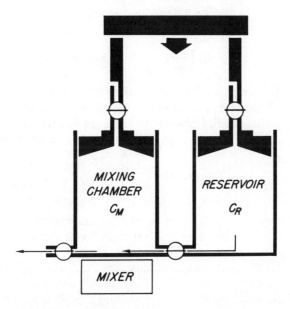

Fig. 4-3 Basic elements of a gradient maker for producing gradients in which density changes *linearly* with volume.

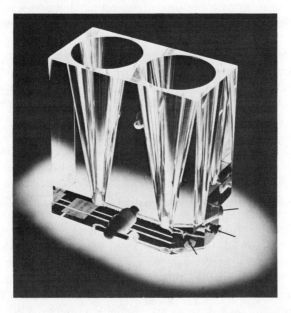

Fig. 4-4 An *open-type* linear gradient maker for centrifuge tubes. (Courtesy of Buchler Instruments, Inc.)

being delivered (i.e., the instantaneous concentration of the mixing chamber) would be

$$C_t = 0 + (40 - 0) \left(\frac{30}{60}\right) = 20 \quad \text{or} \quad 20\% \text{ w/w sucrose}$$

Example 2 Suppose that the mixing chamber contains 15 ml of 18% w/w sucrose and the reservoir contains 15 ml of 2% w/w sucrose (i.e., a linearly decreasing density gradient is being produced). Then at that point in time when 20 ml has been delivered, the concentration being delivered would be

$$C_t = 18 + (2 - 18) \left(\frac{20}{30}\right) = 7.3 \quad \text{or} \quad 7.3\% \text{ w/w sucrose}$$

Linear gradients are simpler in concept than are exponential gradients or gradients that have other profiles, but truly linear density gradients are, in fact, difficult to produce. To be linear, the following conditions must be satisfied: (1) the two chambers must have the *same geometry;* (2) the two chambers must contain the *same volumes at all times;* and (3) the volumes of the two chambers *must diminish at exactly the same rate.* Criterion 1 is readily satisfied, but criteria 2 and 3 are not so easily met. Differences in the density and the viscosity of the gradient limits impose problems of uniform chamber unloading and instantaneous mixing. Even though the chambers have the same shapes and contain the same volumes, the chamber containing the denser liquid exerts more pressure in the line connecting the two chambers together. In "open-type" devices, the pressure difference demands that the chamber containing the lighter solution have a greater volume than the chamber containing the denser solution. For example, 20 ml of a solution of density 1.3 is balanced by a solution of density 1.0, if the volume of the latter is 26 ml (i.e., 20 × 1.3 = 26 × 1.0). Moreover, the difference in the chamber volumes will continuously change as the contents of the chambers are depleted. In the above example, a 46-ml gradient would be produced that has density limits of 1.0 and 1.3, *but the profile will not vary linearly between the two limits.* The creation of a vortex in the mixing chamber raises the effective height of the liquid column, thereby raising the pressure exerted by that column.

   The difficulties just described are resolved when devices of the "closed type" are used. In a closed system, movable pistons are inserted into the cylindrical chambers of the gradient maker and are coupled to a horizontal bar that ensures that the volumes in each chamber remain equal at all times—even though the volumes are continuously diminishing (Fig. 4-3). Several such devices are available commercially. In the gradient maker shown in Fig. 4-5, the face of each piston is tapered and

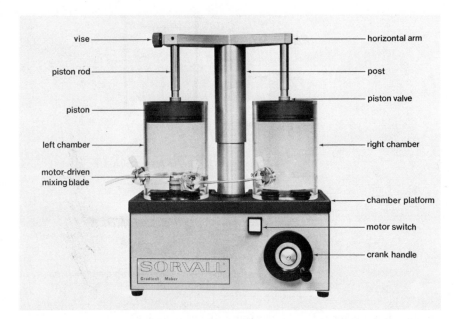

Fig. 4-5 A *closed-type* linear and exponential gradient maker. (Courtesy of E. I. DuPont and Co.)

contains a central opening that conducts air upward through a channel in the piston rod and out through a valve. Consequently, air can be purged from the chambers as the pistons are lowered onto the liquid surfaces, and no vortex is created in the mixing chamber during operation. Gradients produced in such a "closed" device are indeed linear.

Exponential Gradients Although linear density gradients are probably the most common type used, many experiments call for *exponential* gradients—that is, gradients in which the density either increases or decreases *logarithmically*. A logarithmically *increasing* gradient is said to be *convex exponential,* and a logarithmically *decreasing* gradient is said to be *concave exponential.*

Exponential gradients are produced by withdrawing liquid from a mixing chamber, the volume of which is kept constant; liquid is drawn into the mixing chamber from a second chamber (reservoir), the volume of which is allowed to diminish (Fig. 4-6). If the mixing chamber initially contains the light limit of the gradient, the resulting gradient profile is convex exponential. If the mixing chamber initially contains the dense limit of the gradient, the resulting profile is concave exponential. The

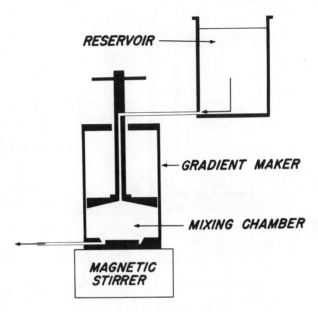

Fig. 4-6 Basic components of an exponential gradient maker.

"degree" of concavity or convexity depends on the selected fixed volume of the mixing chamber and the total volume of gradient to be produced (i.e., the total reservoir volume used). Figures 4-7a and 4-7b show the shapes of 15 convex and concave exponential gradients produced by using fixed mixing chamber volumes of 100 to 1500 ml and total delivered volumes of up to 1500 ml. These families of curves were generated from the relationship

$$C_t = C_R - (C_R - C_M) e^{-V_t/V_M} \qquad (4\text{-}2)$$

which defines the concentration (C_t) of the stream of gradient emerging from the mixing chamber at any time t. In this equation C_R is the concentration of the reservoir liquid (note that this "limit" is never actually reached), C_M is the starting concentration of the mixing chamber, e is the natural base, V_t is the volume of gradient already withdrawn at time t, and V_M is the volume of the mixing chamber.

The curves shown in Fig. 4-7 were designed for use in selecting exponential gradients for large-capacity rotors (e.g., zonal rotors) or for parallel filling of a multiplicity of smaller tubes. However, the same curves can be used for selecting the profiles of smaller gradient volumes simply by dividing the ordinate and fixed volume values by a constant. For example, the 300-ml fixed-volume curve(s) also describes a 30-ml fixed-volume curve(s), so long as the ordinate values are also divided by 10. The following examples illustrate the use of equation 4-2.

Example 1 Suppose that the mixing chamber contains 300 ml of 0% w/w sucrose and the reservoir contains 40% w/w sucrose (i.e., a convex exponential gradient is being generated). Then at that point in time when a volume of 600 ml of gradient has been delivered, the concentration being delivered (i.e., the instantaneous concentration of the mixing chamber)

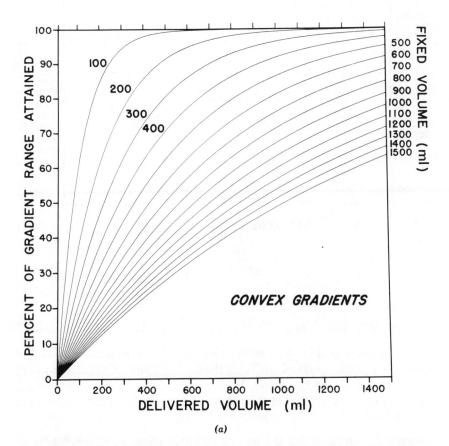

(a)

Fig. 4-7 Profiles of *convex* (a) and *concave* (b) exponential density gradients produced by using specific fixed volumes. See text for explanation and examples of how to use the curves.

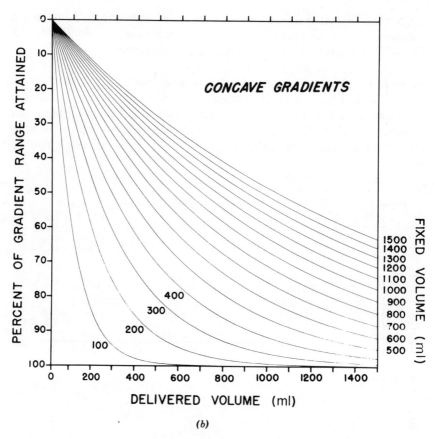

CONCAVE GRADIENTS

(b)

Fig. 4-7 continued

would be

$$C_t = 40 - (40 - 0)\, e^{-600/300} = 34.6 \qquad \text{or} \quad 34.6\% \text{ w/w sucrose}$$

An examination of Fig. 4-7a shows that for a fixed volume of 300 ml, a delivered volume of 600 ml will bring the mixing chamber volume to about 86% of the reservoir limit, and this corresponds to 34.6% w/w sucrose. Clearly, the concentration of the solution being delivered can be obtained by using either Fig. 4-7 or equation 4-2.

Example 2 Suppose that the mixing chamber contains 50 ml of 12% w/w sucrose and the reservoir contains 50% w/w sucrose. Then at that point in time when a volume of 85 ml of gradient has been delivered, the concentration being delivered (i.e., the instantaneous concentration of the mixing chamber) would be

$$C_t = 50 - (50 - 12)e^{-85/50} = 43.1 \quad \text{or} \quad 43.1\% \text{ w/w sucrose}$$

Using Fig. 4-7a to find the answer, for a 50 (or 500) ml mixing chamber volume when 85 (or 850) ml of gradient has been delivered, the percent of the gradient range attained is 81.7. The range is from 12 to 50% or 38 (i.e., 50 minus 12); 81.7% of 38 is 31.1, and 12% plus 31.1% is 43.1% w/w sucrose!

Example 3 Suppose that you wish to prepare a 50-ml concave exponential gradient varying from 40% w/w sucrose to 10% w/w sucrose by using a 20-ml mixing volume. What must the reservoir sucrose concentration be? Using equation 4-2,

$$10 = C_R - (C_R - 40)e^{-50/20}$$

Therefore,

$$C_R = 7.3\% \text{ w/w sucrose}$$

The curves in Fig. 4-7b are used to answer the same question in the following way. The percent of the gradient range attained at 50 (or 500) ml for a 20 (or 200) ml fixed volume is 91.7%. Therefore, the desired range of 40 to 10% can be achieved only if C_R and C_M differ by 32.7% [i.e., $(40 - 10)/91.7\%$]; hence C_R must be 7.3% w/w sucrose (i.e., 40 − 32.7).

Example 4 Suppose that you want to produce a density gradient that has a convex profile like that of the 700-ml fixed-volume curve in Fig. 4-7a. You want a 1000-ml gradient that starts at 8% w/w sucrose and finishes at 52% w/w sucrose. What should the reservoir concentration be? Using equation 4-2,

$$52 = C_R - (C_R - 8)e^{-1000/700} = 65.5\% \text{ w/w sucrose}$$

In other words, to reach a concentration of 52% w/w sucrose after delivering 1000 ml of gradient, the reservoir concentration would have to be 65.5% w/w sucrose. Figure 4-7a may be used to obtain the same conclusion. At a delivered volume of 1000 ml, a 700-ml fixed volume yields 76.5% of the gradient range. Consequently,

$$(C_R - 8)(0.765) = 52 - 8 = 44$$

Therefore

$$C_R = 65.5\% \text{ w/w sucrose}$$

Either equation 4-2 or the sets of curves in Fig. 4-7 can be used as a guide to preparing convex and concave exponential gradients.

Exponential gradient makers are generally simpler in construction than are linear gradient makers because the volume of the mixing chamber is held constant and the respective shapes of the mixing chamber and reservoir are immaterial. It is advantageous that the capacity of the mixing chamber be adjustable so that exponential gradients of varying degree of convexity and concavity can be produced, and it is also desirable to be able to purge air bubbles from the mixing chamber after it has been loaded with the fixed volume. The apparatus shown in Fig. 4-5 can be used to produce exponential gradients by lowering the mixing chamber piston onto the surface of the fixed volume and locking it in position using the vise; the piston is removed from the other chamber, which serves as the reservoir. A simple apparatus used in the author's laboratory is shown in Fig. 4-8.

Variable-Profile Gradients   Some workers find it useful to tailor gradients of complex shape to effect the separation of a given mixture of particles. Such gradients may contain linear and/or exponential regions (which could be produced by the methods described above) but often take more complex forms. The production of such density gradients requires a special gradient generator capable of mixing the initial and final gradients limits in continuously varying proportions. Such a device is shown in Fig. 4-9, in which a slowly revolving cam cut to assume the gradient profile is used as a guide for proportioning the limiting solutions.

Pumps   Density gradients may be delivered to (or removed from) centrifuge tubes in a variety of ways. The simplest method is to mount the gradient maker above the tubes (or rotor) and allow the gradient stream to be delivered by gravity. However, if a pump is used, greater control can be exercised over the rate and the uniformity of the flow. Basically, there are three types of pump: *impeller, piston,* and *peristaltic*. In an impeller-type pump, the rapidly rotating impeller draws the liquid into the pump chamber through one port and expels it through another. In piston pumps the cyclical movement of the piston in and out of a cylindrical chamber alternately fills and empties the chamber through ports controlled by one-way valves. Neither the impeller nor the piston pump is particularly suitable for density gradient work because of the sizable "dead volume" of the pump chambers. Moreover, impeller pumps frequently create turbulence in the gradient stream, and piston pumps deliver

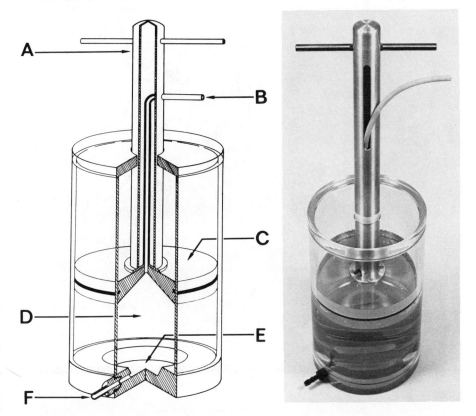

Fig. 4-8 Variable-volume mixing chamber used in the author's laboratory for producing convex and concave density gradients. The desired fixed volume is placed in the chamber and the tapered piston lowered onto the liquid's surface to expel air from the chamber and the inlet line. *A*, piston handle; *B*, inlet from reservoir; *C*, tapered piston; *D*, variable-volume chamber; *E*, recess for stirring bar; *F*, outlet to rotor or tubes.

the "dead volume" in pulses, thereby introducing numerous small steps into the gradient.

Peristalic pumps are preferred to other types because of their greatly reduced dead volume and generally smoother flow rates. In peristaltic pumps the liquid is pushed through a continuous length of small-bore, flexible tubing by the squeezing action of a series of rollers. Since the gradient is always confined to the pump tubing, no contact is made with the metal or plastic parts of the pump. The length of tubing used and its

Fig. 4-9 Beckman Model 141 variable-profile density gradient maker. (Courtesy of Beckman Instruments, Inc.)

diameter are kept small to reduce laminar mixing of the column of gradient as it flows to (or from) the centrifuge tubes. A number of different peristaltic pumps are available commercially (Fig. 4-10), some having reversible motors with variable speed control and accepting multiple heads. These features make them especially desirable for density gradient work. Some commercial gradient makers include a pumping unit (see Fig. 4-9).

COLLECTING FRACTIONS

At the conclusion of centrifugation, the density gradient and the entrained particle zones are usually removed from the tube (or rotor) and collected as a series of fractions. This may be achieved in several ways. One of the oldest and simplest methods is to puncture the bottom of each centrifuge tube and allow the contents to slowly drip out. The gradient may also be removed by carefully lowering a narrow cannula to the bottom of the tube and withdrawing the gradient by use of a peristaltic pump. In both pro-

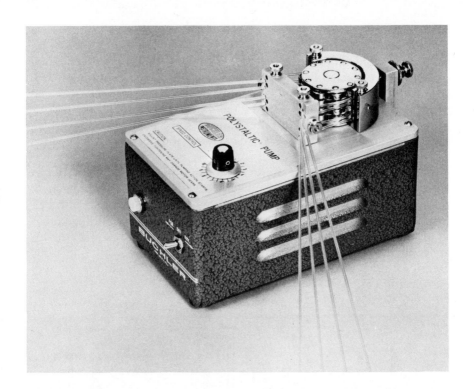

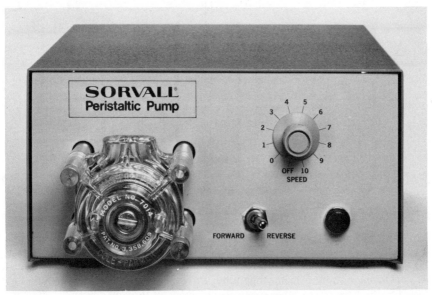

Fig. 4-10 Peristaltic pumps used with density-gradient makers. (Top, courtesy of Buchler Instruments, Inc.; bottom, courtesy of E. I. DuPont and Company)

cedures the gradient exits the tube dense end first, and thus the separated particles are collected in order of decreasing sedimentation rate or density. Another popular method is to insert a conical cap into the open end of the centrifuge tube and displace the gradient up through the cap by introducing a dense solution to the bottom of the tube; in this approach the gradient is collected light end first. The various methods are depicted in Fig. 4-11.

If the temperature of the density gradient differs appreciably from its surroundings, *thermal convection* will occur because different regions of the gradient change temperature at different rates. This can disturb delicate separations but may be avoided by maintaining the centrifuge tubes in a water bath during the unloading operation. A handy apparatus for collecting density gradients through either the top or the bottom of the centrifuge tube while at the same time maintaining constant temperature is shown in Fig. 4-12.

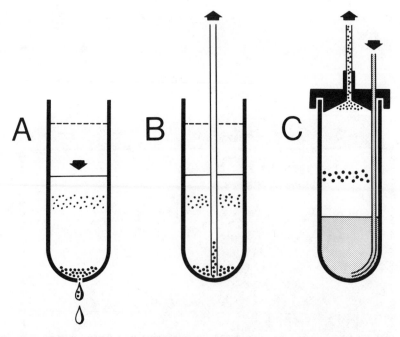

Fig. 4-11 Methods for collecting density-gradient fractions after centrifugation. *A*, puncturing bottom of centrifuge tube; *B*, withdrawing through a cannula inserted to the base of the tube; *C*, displacing gradient through a conical tube cap by delivering a very dense liquid to the base of the tube by using a cannula.

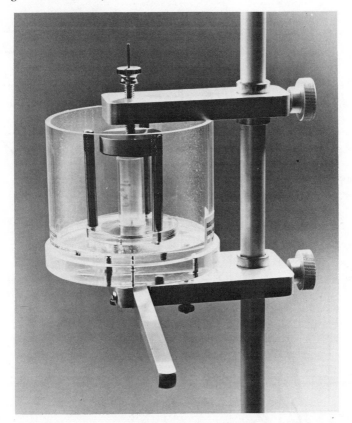

Fig. 4-12 Device for collecting density gradient through either the top or the bottom of a centrifuge tube that can be maintained at a constant temperature in a water jacket. (Courtesy of Buchler Instruments, Inc.)

MEASURING GRADIENT DENSITY

Since the density of a solution is its mass:volume ratio, measurement of these two parameters in successive aliquots of the collected gradient reveals the gradient's profile. However, this approach is awkward and time consuming and is rarely used in connection with density gradient centrifugation. Instead, the *refractive index* of small drops of liquid taken from different regions of the gradient is determined, and the corresponding densities obtained from tables or curves relating refractive index to density (Fig. 4-13).

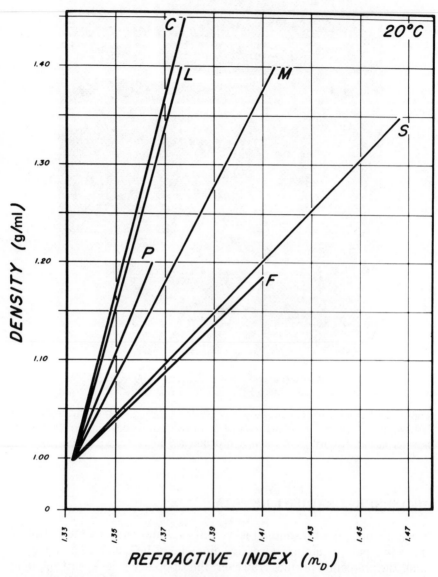

Fig. 4-13 Refractive indeces and densities of a number of solutions that are frequently used in density gradient centrifugation. *C*, CsCl; *L*, Ludox; *P*, Percoll; *M*, metrizamide; *S*, sucrose; *F*, Ficoll.

LIGHT REFRACTION AND SNEL'S LAW

The speed of light changes as light passes from one medium into another. If the light passes across the interface between the two media at any angle other than 90°, the *direction* of the light is also changed (Fig. 4-14). The refractive index of a substance is a measure of its ability to bend the light and is related to the substance's density. Willebrod Snel (1580–1623) showed that the angle at which a light ray is refracted (i.e., angle R) depends on the angle of incidence (i.e., angle I) and the refractive indices of the two substances through which the light travels; that is,

$$n_{D_1} (\sin I) = n_{D_2} (\sin R), \tag{4-3}$$

where n_{D_1} and n_{D_2} are the refractive indices of the two substances and the angles of incidence and refraction are measured relative to a normal to the interface between the two substances. Equation 4-3 is known as Snel's law.

A *refractometer* is an instrument capable of measuring the refractive index of a substance and is an indispensable tool in density gradient centrifugation. The invention of the refractometer, an example of which is shown in Fig. 4-15, is generally credited to Ernst Abbe (1840–1905) in 1874. In the typical refractometer (Fig. 4-16), a beam of polychromatic light is directed through a small drop of sample spread between the sur-

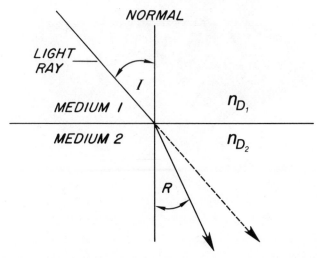

Fig. 4-14 Refraction of light ray (arrow) as it passes across the interface between one medium and another. In this instance the density of medium 2 is greater than that of medium 1, so that the light ray is bent toward the normal.

faces of an *illuminating* (or *diffusing*) prism and a *refracting* prism. Light rays refracted by the sample and the prisms are then reflected off the surface of a *pivot*-mounted *reflector* toward and through a second set of prisms that compensate for chromatic abberation and dispersion. Adjustment of these prisms is monitored visually through the eyepiece of the instrument. The pivot angle of the reflector must be reset for samples of different refractivity, and adjustment of the pivot simultaneously moves a *scale* on which the refractivity may then be read. The refractive index of a solution varies with temperature and it is therefore necessary to maintain a constant temperature when recording the refractive indices of a series of samples. In the refractometer shown in Fig. 4-15, the diffusing

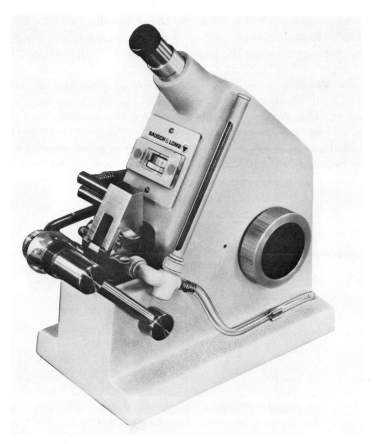

Fig. 4-15 A refractometer (Abbe 4L). Water may be circulated through the housings of the illuminating and refracting prisms so that the sample can be measured at a specific temperature. (Courtesy of Bausch and Lomb.)

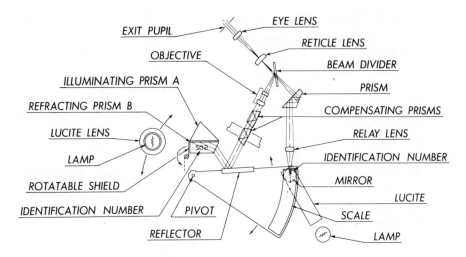

EXIT PUPIL

EYE LENS

RETICLE LENS

OBJECTIVE

BEAM DIVIDER

ILLUMINATING PRISM A

PRISM

REFRACTING PRISM B

COMPENSATING PRISMS

LUCITE LENS

RELAY LENS

LAMP

IDENTIFICATION NUMBER

ROTATABLE SHIELD

MIRROR

LUCITE

IDENTIFICATION NUMBER   PIVOT

SCALE

REFLECTOR

LAMP

Fig. 4-16 Paths of light and basic components of a refractometer. (Courtesy of Bausch and Lomb.)

and refracting prisms are mounted in water-jacketed housings through which water at a constant temperature can be circulated. Figure 4-13 shows the relationship at 20°C between refractive index and concentration for solutions of a number of solutes regularly used in density gradient centrifugation.

GRADIENT MATERIALS

For most biological work, the ideal solute used to produce a density gradient should satisfy the following criteria:

1 The solute should be sufficiently soluble in aqueous media to produce the range of densities needed.
2 Dissolution of the solute should not result in solutions of high viscosity in the desired density range. Viscous solutions pose special mixing problems for gradient makers and are handled with some difficulty. Moreover, the sedimentation rate of a particle decreases directly in relation to the viscosity of the surrounding medium.
3 The solute should not exert too high an osmotic pressure in the desired density gradient range. The size, the shape, and the density of whole cells, membrane-enclosed subcellular organelles, and even macro-

molecules can be altered as these particles sediment through a density gradient if the osmotic pressure of the surrounding medium changes markedly.

4   Solutions of the gradient solute should be adjustable to the pH and ionic strengths that are compatible with the particles being separated.

5   The solute should not react with the particles being separated and should not interfere with the methods of analysis of the collected fractions.

6   Solutions of the gradient solute should exhibit a property that can conveniently be used as a measure of concentration or gradient density.

7   The solute should be readily removable from the collected gradient fractions (should that be necessary or desirable).

8   The solute should be affordable in the quantities needed.

Although certain of these criteria are readily met by most solutes used for density gradient centrifugation, no single solute meets *all* criteria. The solutions formed by most density gradient solutes can be adjusted to the desired pH and ionic strength, and densities can usually be reliably determined from refractive indices of collected fractions. Critical decisions concerning the choice of gradient solute are more frequently based on the relative importance of the other criteria in a given experimental situation. The solutes used to produce aqueous density gradients for centrifugation typically fall into one of the following categories: (1) salts, (2) small, organic molecules, (3) polymers or high-molecular-weight substances, and (4) finely divided, inert particles.

Salts   Over the years, a large number of different salts have been used to produce density gradients, including sodium bromide, sodium iodide, cesium chloride, cesium bromide, cesium sulfate, cesium formate, cesium acetate, rubidium bromide, and rubidium chloride. Among these, cesium chloride (CsCl) has found the most widespread acceptance and use. Cesium chloride possesses a number of properties that make its use especially desirable in many different kinds of density gradient work. Cesium chloride is extremely soluble in water, forming solutions of densities up to 1.92gm/ml (Fig. 4-17). The viscosity of CsCl solutions is very low and actually *diminishes* with increasing densities of up to about 1.4 gm/ml (Fig. 4-18). The densities of CsCl solutions are readily determined by refractive index measurement. Cesium chloride density gradients have proven especially useful for separations of viruses and dense, cellular macromolecules such as DNA, RNA, glycogen, starch, and other poly-

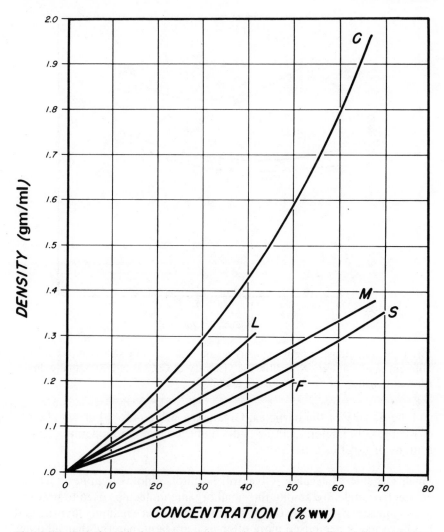

Fig. 4-17 Relationship between concentration and density for various solutions frequently used in density gradient centrifugation. *C*, CsCl; *L*, Ludox or Percoll; *M*, metrizamide; *S*, sucrose; *F*, Ficoll.

saccharides. Among the disadvantages of CsCl are its high ionic strength and osmolarity (Fig. 4-19). High concentrations of CsCl tend to disrupt certain intra- and intermolecular bonds present in proteins and nucleic acids. Because it is hydrophilic, CsCl can compete with hydrated macromolecules and particles for their bound water, thereby altering the size

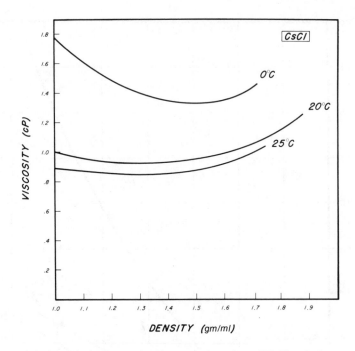

Fig. 4-18   Relationship between density and viscosity of CsCl solutions at three different temperatures. Note how the viscosity of CsCl solutions actually diminishes at densities of up to about 1.4 gm/ml.

and the density of the particles. Organelles, which often behave like microscopic osmometers, can be badly damaged even in moderate concentrations of CsCl or other salts.

Small Organic Molecules   Glycerol, sorbitol, chloral hydrate, sucrose, and metrizamide are among the small organic molecules used to produce density gradients. Among these, sucrose is the overwhelming favorite and has most likely been used more often as a gradient material than all other solutes combined. Sucrose is relatively inexpensive and extremely soluble in aqueous media and can be used to produce density gradients ranging up to 1.35 gm/ml, which includes the density values for most cells and intracellular organelles. Above a density of 1.15 gm/ml, the viscosity of sucrose solutions dramatically rises (Fig. 4-20), thus markedly reducing particle sedimentation rates. Generally, sucrose does not damage cells, organelles, or subcellular particles, although contaminants, including enzymes present in many commercial sources of sucrose, can be deleterious. Although the osmotic pressure of sucrose solutions does not rise as rapidly

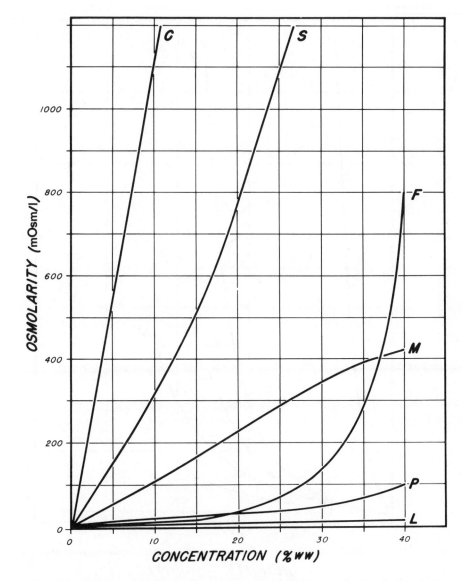

Fig. 4-19 Relationship between concentration and osmolarity for various solutions frequently used in density-gradient centrifugation. *C*, CsCl; *S*, sucrose; *F*, Ficoll; *M*, metrizamide; *P*, Percoll; *L*, Ludox.

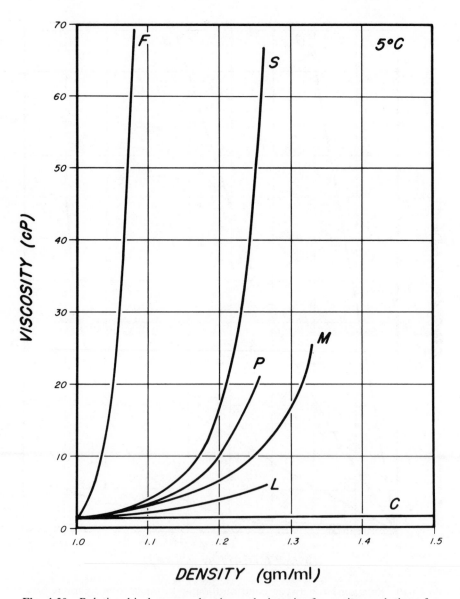

Fig. 4-20 Relationship between density and viscosity for various solutions frequently used in density gradient centrifugation. *F*, Ficoll; *S*, sucrose; *P*, Percoll; *M*, metrizamide; *L*, Ludox; *C*, CsCl.

at increasing solute concentrations as do solutions of CsCl and other salts, the high osmotic pressure of dense sucrose solutions does cause shrinkage of certain cells and organelles. Sucrose has also been shown to interfere in varying degrees with some assay procedures conventionally used to measure marker enzyme activities. Notwithstanding these limitations, sucrose remains an extremely popular gradient solute, and most density gradient fractionation procedures are first attempted by using sucrose, with other gradient solutes turned to when the use of sucrose creates difficulties.

A relative newcomer in this category of gradient solutes is a substance called *metrizamide* [i.e., 2-(3-acetamido-5-*N*-methylacetamido-2, 4, 6-triiodobenzamido)-2-deoxy-D-glucose). Metrizamide is a nonionic, iodinated derivative of glucose initially developed for use as an X-ray contrast medium. Metrizamide is less viscous and exerts less osmotic pressure than does sucrose and can be used to produce density gradients ranging up to 1.4 gm/ml. In comparison with sucrose solutions of the same solute concentrations, metrizamide is denser.

Polymers The greatly reduced osmolarity produced by high-molecular-weight substances even in high concentrations was early recognized as an advantage for density gradient separations of whole cells and membrane-enclosed organelles. Among the high-molecular-weight materials first used for density gradient separations were natural substances such as *glycogen* and plasma proteins (especially *albumin*). Glycogen, a branched polymer of glucose, is quite polydisperse, and the larger molecules of this polysaccharide are readily sedimented, even at modest centrifugal forces. For this and other practical reasons, glycogen is rarely used today, whereas gradients formed either from redissolved plasma proteins or from whole serum (usually calf serum) are still popular, especially for separations of erythrocytes, leukocytes, and other mammalian tissue cells. Among the synthetic polymers, currently the most popular in use is *Ficoll,* an inert material produced by cross-linking sucrose and epichlorohydrin. Ficoll has an average molecular weight of 400,000 and exerts much less osmotic pressure than do sucrose, metrizamide, and salts like CsCl. Gradients that have densities of up to about 1.23 gm/ml can be generated with Ficoll. Probably the most significant disadvantage of Ficoll is its high viscosity, even at low concentrations. Among the polymers used less frequently as density gradient solutes are *dextran* (a polymer of glucose) and *polyvinylpyrrolidone*.

Finely Divided, Inert Particles The most widely used gradient materials belonging to this category are the several forms of *colloidal silica* (also called *silica sol*), such as *Ludox HS, Ludox SM,* and *Percoll.* Colloidal silica takes the form of finely divided, microscopic, spherical particles in which great numbers of atoms of silicon interconnected by oxygen bridges form a three-dimensional meshlike structure. The silica particles are polydisperse; for example, in Ludox HS particle diameters vary from 80 to 250 Å (average diameter is 150 Å), and in Ludox SM the diameters vary from 50 to 150 Å (average diameter is 70 Å). Percoll has been introduced as a density gradient material only recently; it consists of colloidal silica particles coated with polyvinylpyrrolidone (PVP). The average diameter of the Percoll particles is 170 Å, with the PVP content representing about 12%. The advantages of the various forms of colloidal silica for gradient generation include low viscosity and low osmolarity despite the reasonably high density of suspensions of these materials. Probably the most serious limitation to the use of silica-based materials for density gradient separations is the tendency of the silica particles themselves to undergo some sedimentation during centrifugation. Consequently, Ludox and Percoll have found their widest applications in centrifugal separations of large particles such as whole cells and the larger organelles, where the required centrifugal force is low and the centrifugation time is short. Even shallow gradients of Ludox and Percoll are opalescent. Although this does not affect particle separations, it can create difficulty if the visual identification of the separated zone positions in the gradient is desired.

Because silica particles are polydisperse, the distribution of particle sizes through the centrifuge tube changes during the course of centrifugation (even at moderate rpm) as the larger silica particles undergo sedimentation. (This is true also when other polydisperse solutes, such as glycogen, are used.) Consequently, even if the colloidal silica is initially uniformly distributed through the centrifuge tube, a density gradient will be created *automatically* during centrifugation, as the sedimentation of silica particles increases their concentration toward the bottom of the tube and decreases their concentration near the top. Under certain conditions, the automatic formation of a density gradient by centrifuging an initially uniform solution can be achieved with other gradient solutes. These *self-generating density gradients* have special value in certain applications and are considered further in Chapter 5.

The relationships between concentration, refractive index, solution density, viscosity, and osmolarity for the more important gradient solutes are compared in Fig. 4-13, 4-17, 4-19, and 4-20. Properties of additional gradient materials are listed in Table 4-1. As a convenient guide to the specific applications and uses of the various gradients, Table 4-2 lists the

Table 4-1 Properties of Materials Used for Preparing Density Gradients

Material	Molecular Weight or Average Particle Diameter	Maximum Density of Aqueous Solution	20% w/w Solution at 25°C		
			Density	Viscosity (cP)	Refractive Index
Cesium chloride	168	1.92	1.17	2	1.350
Sodium bromide	103	1.51	1.17	2	1.363
Glycerol	92	1.26	1.05	4	1.356
Sucrose	342	1.35	1.08	2	1.363
Ficoll-400^a	400,000	1.23	1.07	27	1.356
Dextran^b	72,000	1.05	1.04^f	5^f	1.348^f
Ludox^c HS	150 Å	1.4	1.13	2	1.347
Ludox^c SM	70 Å	1.4	1.13	2	1.347
Percoll^d	170 Å	1.2	1.13	8	1.354
Metrizamide^e	789	1.4	1.10	2	1.362

^{a} A highly branched copolymer of sucrose and epichlorohydrin.
^{b} Branched chains of α-1-6 linked glucosyl residues.
^{c} Microscopic, spherical particles of silica, forming colloidal suspension in water.
^{d} Microscopic spherical particles of silica covered with polyvinylpyrrolidone, forming colloidal suspension in water.
^{e} 2-(3-Acetamido-5-N-methylacetamido-2,4,6-triiodobenzamido)-2-deoxy-D-glucose.
^{f} Dextran cannot be prepared at 20% w/w; the density, the viscosity, and the refractive index given are for a 10% w/w solution.

types of biological particle that have been effectively separated using CsCl, sucrose, metrizamide, Ficoll, Ludox, and Percoll.

A final point concerning the properties of gradient solutes is that certain properties exhibited by an aqueous solution of the solute may be altered by the dissolution of other materials in the gradient. Needless to say, a 10% w/w solution of Ficoll prepared in physiological saline has a slightly different density, refractive index, osmolarity, and so on than does 10% w/w Ficoll in water alone (see Appendix for physical properties of gradient solutes in water and in saline). With finely divided particles, the effect of addition of even small quantities of ions or other substances to the solution may alter these parameters more extensively, markedly changing viscosity and osmolarity. For example, the addition of small quantities of salts to concentrated Ludox solutions is soon followed by transformation of the solution to an intractable gel.

Table 4-2 Density Gradients Used to Separate Cells, Subcellular Particles, Viruses, and Macromolecules

Particle	Gradient Material					
	CsCl	Sucrose	Metrizamide	Ficoll	Ludox	Percoll
Whole cells		+	+	+	+	+
Nuclei		+	+	+	+	
Chloroplasts		+		+	+	
Mitochondria		+		+	+	
Lysosomes		+		+	+	+
Peroxisomes and glyoxysomes		+		+	+	
Plasma membranes		+			+	+
Ribosomes and ploysomes	+	+	+			
Golgi bodies		+				
Viruses	+	+			+	+
Polysaccharides	+		+			
Proteins	+		+			
Nucleic acids	+	+	+			
Lipids			+			

REFERENCES AND RELATED READING

Books

Behrens, M., Zell- und Gewebetrennung. In *Handbuch der Biologischen Arbwitsmethoden* Vol. 5, (E. Abderhalden, Ed.) Urban and Schwarzenberg, Berlin, 1938, Part 10, 1363.

Birnie, G. D., and Rickwood, D. Ed. *Centrifugal Separations in Molecular and Cell Biology.* Butterworths, London, 1978.

Catsimpoolas, N. *Methods of Cell Separation.* Plenum, New York, 1977.

Hinton, R., and Dobrota, M. *Density Gradient Centrifugation.* North-Holland, Amsterdam, 1976.

Liteanu, C., and Gocan, S. *Gradient Liquid Chromatography.* Wiley, New York, 1974.

Pertoft, H., and Laurent, T. C. The use of gradients of colloidal silica for the separation of cells and subcellular particles. In *Modern Separation Methods of Macromolecules and Particles,* T. Gerritsen, Ed. Wiley-Interscience, New York, 1969.

Rickwood, D., Ed. *Biological Separation in Iodinated Density-Gradient Media.* Information Retrieval, London, 1976.

Rickwood, D., Ed. *Centrifugation: A Practical Approach*. Information Retrieval, London, 1978.

Articles and Reviews

Anderson, N. G. (1955) Mechanical device for producing density gradients in liquids. *Rev. Sci. Instrum., 26*, 891.

Anderson N. G. (1955) Studies on isolated cell components. VIII. High resolution gradient differential centrifugation. *Exp. Cell Res., 9*, 446.

Anderson, N. G., and Rutenberg, E. (1967) Analytical techniques for cell fractions. VII. A simple gradient-forming apparatus. *Anal. Biochem., 21*, 259.

Bell, L. R., and Hsu, H. W. (1974) Transport phenomena in zonal centrifuge rotors. IX. Gradient properties of Ficoll and methyl cellulose. *Separation Sci., 9*, 401.

Birnie, G. D., and Harvey, D. R. (1968) A simple density-gradient engine for loading large-capacity zonal ultracentrifuge rotors. *Anal. Biochem., 22*, 171.

Brakke, M. K. (1951) Density gradient centrifugation: A new separation technique. *J. Amer. Chem. Soc. 53*, 1847.

Brakke, M. K. (1979) The origins of density gradient centrifugation. *Fractions*, No. 1, 1.

Choules, G. L. (1962) A linear-gradient mixing device for viscous solutions. *Anal. Biochem., 3*, 236.

de Duve, C. (1964) Principles of tissue fractionation. *J. Theoret. Biol., 6*, 33.

de Duve, C., Berthet, J., and Beaufay, H. (1959) Gradient centrifugation of cell particles. Theory and applications. *Progress Biophys. Biophys. Chem., 9*, 325.

Hinton, R. H., and Dobrota, M. (1969), A simple gradient maker for use with zonal rotors. *Anal. Biochem., 30*, 99–110.

Kahler, H. J., and Lloyd, B. J. (1951) Sedimentation of polystyrene latex in a swinging-tube rotor. *J. Phys. Colloid Chem., 55*, 1344.

Lyttleton, J. W. (1970) Use of colloidal silica in density gradients to separate intact chloroplasts. *Anal. Biochem., 38*, 277.

Mathias, A. P., and Wynter, C. V. A. (1973) The use of metrizamide in the fractionation of nuclei from brain and liver tissue by zonal centrifugation. *FEBS Lett. 33*, 18.

Moore, D. H. Gradient centrifugation. In *Physical Techniques in Biological Research*, Vol. II, Part B, D. H. Moore, Ed. Academic, New York, 1969.

Munthe-Kaas, A. C., and Seglen, P. O. (1974) The use of metrizamide as a gradient medium for isopycnic separation of rat liver cells. *FEBS Lett. 43*, 252.

Pertoft, H., Laurent, T. C., Laas, T., and Kagedal, L. (1978) Density gradients prepared from colloidal silica particles coated by polyvinylpyrrolidone (Percoll). *Anal. Biochem., 88*, 271.

Pickels, E. G. (1943) Sedimentation in the angle centrifuge. *J. Gen. Physiol.*, **26**, 341.

Pretlow, T. G., Boone, C. W., and Shrager, C. W. (1969) Rate zonal centrifugation in a Ficoll gradient. *Anal. Biochem.*, **29**, 230.

Rickwood, D., and Birnie, G. D. (1975) Metrizamide, a new density gradient medium. *FEBS Lett.* **50**, 102.

Literature Available from Commercial Distributors of Gradient Materials
Dextran. Pharmacia Fine Chemicals AB, Box 175, S-751 04 Uppsala 1, Sweden.

Ficoll For Cell Research. Pharmacia Fine Chemicals AB, Box 175, S-751 04, Uppsala 1, Sweden.

Metrizamide, A Gradient Medium for Centrifugation Studies. Accurate Chemical and Scientific Corporation, 28 Tec St., Hicksville, N. Y. 11801, USA.

Percoll for Density Gradient Centrifugation. Pharmacia Fine Chemicals AB, Box 175, S-751 04, Uppsala 1, Sweden.

FIVE

Density Gradient Centrifugation

RATE AND ISOPYCNIC SEPARATIONS OF PARTICLES

Rate and *isopycnic* centrifugation are two alternative forms of the density gradient approach to the separation of mixtures of particles. In this chapter we consider the principles and the characteristics of both of these forms and the criteria one uses, given a particular experimental situation, for selecting the most effective method to use. To begin, let us reexamine equation 2-32, which is reproduced here for convenience:

$$\frac{dx}{dt} = \frac{2r^2 \, (\rho_P - \rho_M) \, \omega^2 x}{9\eta \, (f/f_0)} \qquad \text{(4-1 and 2-32)}$$

This equation, derived in Chapter 2, describes the instantaneous rate of sedimentation of particles through a medium of specific density and viscosity. It is seen that if the angular velocity (ω) of the centrifuge rotor remains constant, the value of dx/dt for any particle increases with increasing distance from the axis of rotation (i.e., with increasing values of x). Although it is derived for particles that sediment through solutions of uniform density and viscosity, equation 4-1 also defines the instantaneous sedimentation rate of particles that move through a density gradient. Clearly, under these conditions, dx/dt varies not only with changing values of x, but also with changing values of ρ_M and η.

Our consideration of rate and isopycnic density gradient centrifugation can be simplified somewhat by making the following two assumptions:

1 It is assumed that we are dealing with *spherical* particles, so that equation 4-1 simplifies to

$$\frac{dx}{dt} = \frac{2r^2(\rho_P - \rho_M)\omega^2 x}{9\eta} \qquad (4\text{-}2)$$

2 We assume that the values of r and ρ_P for a particle do not change as it sediments through the gradient. Some biological particles behave like miniature osmometers, so that some change in their volume (usually a *decrease*) does occur as the particles sediment deeper into the density gradient. Consequently, their sedimentation is accompanied by a small change in r and ρ_P. However, for most biological particles, gradient solutes and density profiles can be selected that have minimal effects on these two parameters.

Of the two particle properties that affect dx/dt, namely, r and ρ_P, r has the greater influence because it is a second-order term (i.e., $dx/dt \sim r^2$). For this reason, it is commonly stated that *particle size more markedly influences sedimentation rate than does particle density*. Whether a particular density gradient separation is classified as rate or isopycnic is determined by (1) the values of r and ρ_P of the sedimenting particles and (2) the range over which ρ_M extends in the density gradient. *True* rate and *true* isopycnic separations represent the extremes of a continuum of possibilities, and most experimental density gradient separations fall somewhere between, incorporating aspects of both forms.

RATE SEPARATIONS

In a true *rate separation*, the maximum value of ρ_M in the density gradient is *smaller* than the values of ρ_P for the particles in the sample to be fractionated. Consequently, for every particle, the value of dx/dt is always greater than zero, and given enough centrifugation time, all particles will sediment to the bottom of the tube. However, *before* any of the particles are pelleted, centrifugation is terminated. Thus the particles are distributed through the density gradient as a series of zones, the positions of which are related to particle sedimentation rates.

As has already been noted, the sedimentation rate of a particle during density gradient centrifugation changes in relation to a number of variables (e.g., distance from the axis of rotation, gradient density, gradient viscosity); therefore, in a given density gradient, a characteristic curve for each particle relates dx/dt to time of centrifugation. As an illustration, Fig. 5-1 shows the sedimentation rates for two hypothetical particles, A and B. For each particle, the sedimentation rate rises to attain a maximum value and then progressively diminishes. In this illustration the sedimentation rate of particle A between time t_0 and t_1 is greater than that of particle B; however, as the density of particle B is the greater, a point in time is eventually reached at which the sedimentation rates of both particles become equal (i.e., at t_1). After this time the sedimentation rate of B is greater. For such a pair of particles, it is clear that they are maximally separated *when their sedimentation rates are equal* (except, perhaps, if the run is to be isopycnic—see below). In the majority of instances it is necessary to attempt the separation of more than just two families of particles, so that the situation becomes more complex than that presented in Fig. 5-1. Nonetheless, an appreciation of the manner in which the relative positions of particle zones change within the density gradient and the dependence on changing values of dx/dt for each family of particles is fundamental to good experimental design.

Because the densities of most biological particles are less than about 1.3 gm/ml, rate separations are necessarily performed in *shallow* density

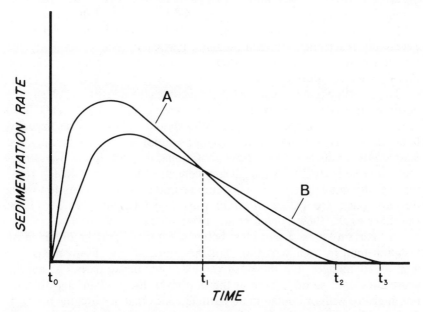

Fig. 5-1 Sedimentation rates through a density gradient of two hypothetical particles, A and B. The particles are maximally separated when their sedimentation rates are equal (at time t_1). See text for further explanation.

gradients. Most biological particles, especially the cellular organelles, are spheroidal and their densities fall in a narrow range; consequently, rate separations of these particles are achieved primarily on the basis of differences in the values of r. Therefore, rate separations are usually thought of as separations based on particle size differences. At the end of the chapter are listed a number of references to articles describing separations carried out on a rate basis.

ISOPYCNIC SEPARATIONS

If the range of the density gradient encompasses the densities of all the particles in the sample being fractionated, and if centrifugation is carried out for a sufficiently long period of time, each particle will sediment through the gradient until it reaches the position where $\rho_M = \rho_P$. Since at that point the term $(\rho_P - \rho_M)$ becomes zero, so does dx/dt and the particles sediment no further. Each particle is said to be at its *banding density* or *isopycnic* point, and continued centrifugation achieves nothing. The final distribution of particle zones in the density gradient is in the order of particle density and is independent of particle sizes or other parameters. For the two particles depicted in Fig. 5-1, if the dense limit of the density gradient is denser than particle B, then both particles will eventually reach their isopycnic points. Particle A reaches its isopycnic position at time t_2, whereas particle B continues to sediment, reaching its isopycnic position at time t_3.

To be certain that a density gradient includes the densities of all sample particles, isopycnic runs are normally carried out in *steep* gradients. Two special aspects of isopycnic separations should be noted. The particles to be separated can be mixed with a dense solution and the mixture layered *beneath* the gradient. Under these conditions, the term $(\rho_P - \rho_M)$ is negative (and so is dx/dt), so that during centrifugation the particles "rise" through the gradient to their isopycnic positions. Experiments following this approach are called *flotation* runs (e.g., Leighton et al., 1968; Fleischer et al., 1969; Nigam et al., 1971; Wilcox et al., 1971).

For certain kinds of particles, better separations may be obtained by flotation than by sedimentation, but the approach has disadvantages as well as advantages. On the positive side, very dense particles, debris, aggregates, and so on remain at the bottom of the gradient and thus do not interfere with or contaminate particle zones that are floating upward. However, a serious disadvantage of the flotation technique is that the sample must be suspended in a sufficiently dense medium to be initially stable at the bottom of the gradient. The elevated viscosity and osmolarity of such a suspending medium may be detrimental to the particles being separated.

Since particles will either sediment or float to their banding densities in an isopycnic gradient, the starting position of the particles is immaterial. That is to say, the particles may even be uniformly distributed through the density gradient at the outset of centrifugation (Fig. 5-2).

Self-Generating Density Gradients If ω is sufficiently high, or if the gradient solute particles are large and polydisperse (e.g., the various forms of colloidal silica), then a *uniformly dense* solution (or colloidal suspension) will be converted to a *density gradient* during centrifugation. For relatively small solute molecules, such as CsCl and metrizamide, the rotor speed must be sufficiently high to cause sedimentation of the solute (Meselson et al., 1957; Flamm et al., 1972; Hell et al., 1974). A stable gradient is formed when the radial sedimentation of the solute is balanced by the centripetal diffusion of molecules along the concentration gradient (i.e., diffusion occurs in a direction opposite to that of sedimentation). The nature of the equilibrium established between sedimentation and diffusion depends on the RCF experienced by the solute particles. Consequently, the shape of the density gradient that is generated is influenced

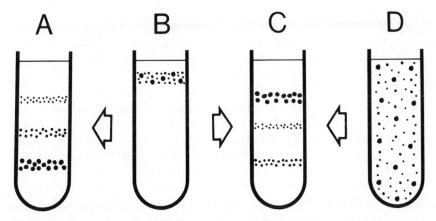

Fig. 5-2 *Rate* and *isopycnic* density gradient separations of particles. In a rate separation (i.e., from *B* to *A*), the particles become distributed in order of their sedimentation coefficients. In isopycnic separations (from *B* to *C* and from *D* to *C*) the particles become distributed in order of their densities. Note that the final distributions of particles (i.e., stages *A* and *C*) may not be the same. Isopycnic separations can sometimes be achieved by initially suspending the particles in a uniform solution of the gradient solute and allowing the gradient to form automatically during centrifugation (i.e., from *D* to *C*).

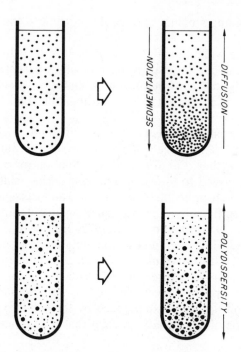

Fig. 5-3 Self-generating density gradients. *Top*, gradient formed as the sedimentation of a monodisperse solute (e.g., CsCl) is opposed by diffusion. *Bottom*, gradient formed by the differential sedimentation rates of a polydisperse solute.

by rotor speed. In addition, it is found that self-generating density gradients are more effectively formed in fixed-angle rotors than in swinging-bucket rotors (Flamm et al., 1972; Pertoft et al., 1978). For large, polydisperse solutes such as Ludox and Percoll, gradients can be formed from uniform suspensions, even at modest rotor speeds, as the centrifugal force redistributes the particles in the tube (Pertoft et al., 1967; Pertoft and Laurent, 1969; Pertoft et al., 1978). These effects are depicted diagrammatically in Fig. 5-3. Self-generating CsCl and metrizamide gradients are especially popular for isopycnic runs, whereas automatically forming Ludox and Percoll gradients are customarily used for rate separations of particles.

COMBINED RATE–ISOPYCNIC SEPARATIONS

More often than not, the separation of particles in a density gradient is the result of *combined* rate and isopycnic effects. That is to say, at the time that centrifugation is halted, some of the particles have reached their isopycnic banding densities whereas others would continue to sediment if the centrifugation time were extended. For example, if a tissue homogenate (e.g., liver) is centrifuged through a 10 to 65% w/w sucrose density gradient for 30 min at $10,000g_{ave}$, subcellular particles such as nuclei and mitochondria will be banded isopycnically, whereas ribosomes, glycogen particles, and fragments of cellular membranes will not yet have reached their isopycnic positions. The latter particles are distributed through the gradient in order of their sedimentation coefficients. Thus, for the cell nuclei and mitochondria, the run is isopycnic, but for the ribosomes, glycogen, and cellular membranes, the run is considered a rate separation.

Prior knowledge of the densities and/or sedimentation coefficients of the various types of particle present in a mixture to be fractionated materially simplifies the design of gradient and centrifugation conditions needed to affect a separation. Figure 5-4 relates the sedimentation coefficients and banding densities of a number of subcellular particles, macromolecules, and viruses; this is commonly known as an "s-ρ" chart. An examination of Fig. 5-4 shows that ribosomal subunits, monosomes, and polysomes cannot be separated by an isopycnic run in a steep gradient because these particles have the same banding density. However, in a shallow density gradient, these particles can be separated on the basis of differences in their sedimentation coefficients. A similar situation exists in attempts to separate smooth endoplasmic reticulum (ribosome-free intracellular membranes), Golgi bodies (also membranous structures), and plasma membranes. Although these particles have closely similar banding densities and are not readily separated isopycnically, they can be separated on a rate basis because of their different sedimentation coefficients.

Figure 5-4 also shows that whereas certain particles may have similar s values and are not readily separated in shallow density gradients on a rate basis, the same particles may be separated in an isopycnic run by using a steep density gradient because the particles have different banding densities. For example, glycogen particles in a liver tissue homogenate are readily separated from smooth and rough (i.e., ribosome-laden) endoplasmic reticulum by isopycnic banding, even though their sedimentation coefficients overlap.

Two-Dimensional Density Gradient Centrifugation   The purity of a fraction obtained by density gradient centrifugation can often be enhanced by carrying out a *two-dimensional* separation (Anderson et al., 1966). In this procedure, a mixture of particles is first separated by rate centrifugation into fractions containing particles of the same or similar s values; following this, the particles in each fraction are subjected to isopycnic

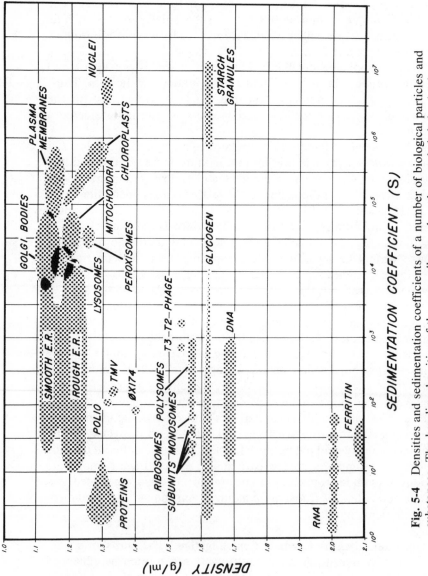

Fig. 5-4 Densities and sedimentation coefficients of a number of biological particles and substances. The banding densities of the organelles are based upon their behavior in sucrose density gradients, whereas the banding densities of viruses and macromolecules are based upon results using CsCl gradients.

separation. As a result, particles of different density that are recovered in the same "rate fraction" now separate isopycnically. Of course, the procedure can also be reversed; that is, the original mixture can be fractionated isopycnically, and then particles in the same "density fraction" can be separated on the basis of differences in their s values.

In a given gradient solute, the banding density of a particle is more or less constant (some particles do exhibit different banding densities in different gradient solutes). In contrast, the value of s changes as the particles encounter increasing gradient densities and viscosities. This fact should be kept in mind when performing rate separations. Two different particles recovered in the same zone following a rate run may, in fact, be separable in the *same* density gradient if either the RCF or the duration of centrifugation is decreased or increased. Thus during a rate run, one zone of particles initially trailing behind another because its s value(s) in the first part of the gradient is lower may later catch and then pass a second zone in which the s values of the particles are now diminishing more rapidly.

At this juncture it is worth repeating a point made earlier (Chapter 3). If the sedimentation coefficients of particles one is attempting to separate differ by an order of magnitude or more (e.g., cell nuclei and mitochondria), the use of a density gradient can be circumvented altogether. In such a situation conventional *differential centrifugation,* followed by successive washings of the desired pellet, may yield entirely satisfactory levels of purity.

SELECTIVE ALTERATION OF PARTICLE BANDING DENSITIES AND SEDIMENTATION COEFFICIENTS

Some particles are especially difficult to isolate because their s and ρ_P values are similar to other particles with which they generally occur. For example, the s and ρ_P values of lysosomes are similar to those of small mitochondria (see Fig. 5-4). In such instances, it is sometimes possible to selectively alter the properties of one of the particles present so that a separation can be achieved more easily. When tissues are treated with dextran or Triton WR 1339, the lysosomes *increase* in size but are *reduced* in density by incorporating quantities of these substances; consequently, the lysosomes can be more easily separated from the mitochondria when homogenates of these tissues are subjected to isopycnic density gradient centrifugation (Leighton et al., 1968; Burge and Hinton, 1970). The addition of $MgCl_2$ to tissue homogenates induces the association of free ribosomes with the endoplasmic reticulum, thus increasing the sedimentation coefficient and the density of the endoplasmic reticulum fragments and making them easier to separate from other constituents of the homogenate (Hinton et al., 1967). Addition of small amounts of Pb^{++} also alters the banding density of endoplasmic reticulum (Hinton et al., 1970). The banding densities of proteins labeled with isotopes of carbon and nitrogen may also be altered to the extent that they may be separated from unlabeled proteins (Hu et al., 1962). Finally, by judiciously selecting the solute to be used in preparing the density gradient as well as the concentration limits, it is possible to selectively alter the sizes and the densities of certain particles in the sample through osmotic effects.

RATE VERSUS ISOPYCNIC SEPARATIONS: ADVANTAGES AND DISADVANTAGES

The choice of whether to try to isolate particles on a rate basis or by isopycnic banding should take into account the following points:

1 For large particles such as cellular organelles, rate runs are of considerably shorter duration than are isopycnic runs. This can be an important consideration where prolonged particle viability or lability are questionable.

2 Since rate runs are usually carried out in shallower density gradients than are isopycnic runs, the sedimenting particles are exposed to lower concentrations of gradient solute, reduced viscosity, and lower osmotic pressure. Such conditions can prove less damaging to particles that are osmotically active.

3 Rate runs are usually carried out by using a lower RCF than is used in isopycnic runs. For certain especially fragile particles, the reduced hydrodynamic shear during sedimentation may be an aid in maintaining particle integrity.

4 The duration of centrifugation is not as critical in an isopycnic run as in a rate run. To attain particle banding, a certain *minimum* amount of centrifugation is required, but extending the centrifugation time beyond this does not alter the results. In contrast, many rate separations can be achieved only by rigorous control of the centrifugation time and RCF, including the time and RCF in effect during rotor *acceleration* and *deceleration* (see Chapter 6). The conditions required to achieve a rate separation usually have to be determined empirically by using a number of trial runs.

5 Generally speaking, greater numbers of particles can be separated by isopycnic density gradient centrifugation than by rate centrifugation. This is because the density gradients used for isopycnic runs are steeper and stably support a greater quantity of sample.

USE OF STEP GRADIENTS

The preparation and the characteristics of step gradients were considered in Chapter 4. It was noted that in most density gradient centrifugation experiments employing step gradients, the gradient is allowed to stand for some time (several hours or more) until the sudden changes in gradient density across each step interface are smoothed out by diffusion. For certain kinds of isopycnic (and sometimes rate) separations, the abrupt density steps that persist for some time are used as surfaces onto which particles can be sedimented during centrifugation, thereby forming discrete particle layers at each step. In such applications it is common to use only a few steps (usually two or three). For example, the centrifuge tube might be filled with three density layers ($\rho_M = 1.1$, $\rho_M = 1.2$, and $\rho_M = 1.3$ in Fig. 5-5) and the mixture of particles to be separated layered onto the top. During centrifugation, each step acts to halt any further sedimentation of particles that have a density lower than the density that occurs at the step. As a result, the sample may be separated into four fractions (see Fig. 5-5). It is important to note that when this technique is used, particles that pass across an interface between steps create an unstable zone (i.e., a region that is denser than the liquid immediately below), with the result that *streaming* and *inversion* quickly carry the

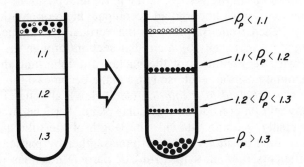

Fig. 5-5 Use of step gradients to selectively halt the sedimentation of families of particles falling within specific density ranges.

particles down to the next interface. Consequently, stable zones are not formed in the regions between successive steps. Step gradients can also be used for flotation experiments, except that in this instance particles rise upward until they encounter an interface above which the liquid density is less than that of the particles.

CHOICE OF SWINGING-BUCKET OR FIXED-ANGLE ROTORS

In most laboratories density gradient centrifugation is performed by use of conventional swinging-bucket and fixed-angle rotors. The more specialized *zonal* and *vertical tube* rotors also are used for density gradient separations, but because of the special nature of these rotors, they are discussed separately in Chapters 6 and 7. Let us consider the bases for choosing between employing a swinging-bucket or fixed-angle rotor for rate or isopycnic separations.

Swinging-Bucket Rotors *Both* rate and isopycnic density gradient centrifugation can be carried out by using swinging-bucket rotors. Typically, the gradient is formed in the centrifuge tubes at rest and the sample carefully layered onto its surface. Often a *cushion* of dense solution is used to support the gradient and prevent particles from entering the hemispherical bottom of the centrifuge tube, and an *overlay* of light solution is placed above the sample to eliminate the meniscus and assure a uniformly thick sample zone.

After the tubes are carefully placed in the rotor buckets, the rotor is slowly accelerated. During acceleration, the buckets progressively reorient (i.e., swing) from the vertical to the horizontal position. The density gradient and the sample undergo an accompanying reorientation. Before reorientation, each *isodense plane* in the vertically positioned gradient is represented geometrically by a circular section through the centrifuge tube. Following reorientation of the bucket, the tube, and the gradient to the horizontal position, each isodense plane becomes a tiny section of a very steep *paraboloid of revolution* (see below), although in practical terms it may still be considered a circular plane. Particles in the sample sediment radially, passing "down" the length of the tube and forming zones whose final positions are determined either by particle sedimentation coefficients (i.e., in a rate run) or particle densities (i.e., in an isopycnic run). Following gradual deceleration of the rotor and reorientation of the bucket, the tube, and the gradient to the vertical position, the separated zones are represented by horizontal layers. Both acceleration and deceleration, especially in the range of 0 to 300 rpm, must be carried out slowly and smoothly in order to minimize Coriolis effects and swirling.

As noted in Chapter 3, even in swinging-bucket rotors, the nonsectorial shape of the conventional centrifuge tube results in convection along the tube wall. The presence of a density gradient materially reduces the effects of this convection with the result that in most instances all particles belonging to a homogeneous population move equal distances down the tube in the same amount of time. However, it is not uncommon to find that at the end of a run the concentration of particles in a given zone is greater at the edges of the zone (i.e., near the tube wall) than at the center.

Fixed-Angle Rotors   The behavior of density gradients and the movements of sedimenting particles are considerably more complicated when density gradient centrifugation is carried out in fixed-angle rotors (Fig. 5-6) than when performed in swinging-bucket rotors (e.g., Fisher et al., 1964; Flamm et al., 1972; Sheeler, 1974). When the gradient and the sample are initially placed in the rotor, each isodense plane takes the form of an ellipse (Fig. 5-6B). Since the tubes are maintained at a constant incline, only the gradient reorients as the rotor is accelerated. Each isodense ellipse is transformed into a small segment of a paraboloid of revolution (Fig. 5-6C) whose focus is on the axis of rotation. The isodense paraboloids become increasingly steep as the rotor gains speed (i.e., as the RCF increases) and eventually approach "verticality" (Fig. 5-6D). At this point the radial distribution of the gradient is similar to that in the tubes of a swinging-bucket rotor.

Convection is far greater in fixed-angle rotors than in swinging-bucket rotors because most of the particles in the sample zone will encounter the tube wall before moving very far through the density gradient (see Chapter 3). Because convection carries particles down the tube along the wall more rapidly than they would travel unimpeded through the gradient, rate separations are rarely attempted in fixed-angle rotors. On the other hand, isopycnic separations are attainable in fixed-angle rotors because particles sedimenting uninterrupted through the gradient as well as those sedimenting along the wall of the tube will come to rest when they reach their banding density (Fig. 5-6E). Indeed, other conditions being the same, the convection that occurs in fixed-angle rotors acts to bring particles to their isopycnic positions more quickly than in swinging-bucket rotors. Other factors also contribute to the high efficiency of fixed-angle rotors for isopycnic separations, including reduced gradient path length, increased cross-sectional area (ellipses vs. circles), and so on; these are taken up in Chapter 8, which examines rotor efficiency. With particle banding com-

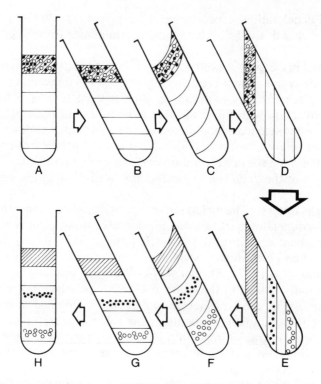

Fig. 5-6 Density gradient centrifugation in a fixed-angle rotor. See text for explanation.

pleted, the rotor is decelerated and the gradient and particle zones reoriented into the vertical position (Figs. 5-6F and 5-6G). Smooth and gradual rotor deceleration is even more crucial to the maintenance of gradient and entrained particle zone stability when using fixed-angle rotors than is the case for swinging buckets. The phenomenon of gradient reorientation and other implications of the "reograd" approach are treated more fully in Chapters 6 and 7 in connection with *zonal* rotors and *vertical tube* rotors.

It was noted earlier (Chapter 3) that, in general, fixed-angle rotors offer increased capacity (i.e., greater numbers of tube positions and/or higher capacity per position and increased RCF in comparison with swinging-bucket rotors. With regard to the preceding discussion, it can be seen that capacity advantages must be weighed against the convection effects that accompany the use of fixed-angle rotors for density gradient separations.

REFERENCES AND RELATED READING

Books

Anderson, N. G., Ed. *The Development of Zonal Centrifuges and Ancillary Systems for Tissue Fractionation and Analysis.* National Cancer Institute Monograph 21. National Cancer Institute, Bethesda, Maryland, 1966.

Birnie, G. D., Ed. *Subcellular Components; Preparation and Fractionation.* Butterworths, London, 1972.

Birnie, G. D., Ed. *Subnuclear Components; Preparation and Fractionation.* Butterworths, London, 1976.

Birnie, G. D., and Rickwood, D., Ed. *Centrifugal Separations in Molecular and Cell Biology.* Butterworths, London, 1978.

Evans, W. H. *Preparation and Characterization of Mammalian Plasma Membranes.* Elsevier-North Holland, Amsterdam, 1978.

Hinton, R., and Dobrota, M. *Density Gradient Centrifugation.* North-Holland, Amsterdam, 1976.

Reid, E., Ed. *Methodological Developments in Biochemistry,* Vol. 4, *Subcellular Studies.* Longmans Group Publishers, London, 1974.

Rickwood, D., Ed. *Biological Separations in Iodinated Density Gradient Media.* Information Retrieval, London, 1976.

Articles and Reviews

Anderson, N. G., Harris, W. W., Barber, A. A., Rankin, C. T., and Candler, E. L. (1966) Separation of subcellular components and viruses by combined rate- and isopycnic-zonal centrifugation. *Nat. Cancer Inst. Monograph,* **21,** 253.

Brakke, M. K. (1979) The origins of density gradient centrifugation. *Fractions,* No. 1, 1.

Burge, M. L. E. and Hinton, R. H. (1970) Isolation of hepatic lysosomes and peroxisomes. In *Advances with Zonal Rotors,* E. Reid Ed. Wolfson Bioanalytical Centre Publishers, London.

Casteneda, M., Sanchez, R., and Santiago, R. (1971) Disadvantages of sucrose-density gradient centrifugation in fixed-angle rotors. *Anal. Biochem.* **44,** 381.

Fisher, W. D., Cline, G. B., and Anderson, N. G. (1964) Density gradient centrifugation in angle-head rotors. *Anal. Biochem.,* **9,** 477.

Flamm, W. G., Birnstiel, M. L., and Walker, P. M. B. (1972) Isopycnic centrifugation of DNA; methods and applications. In *Subcellular Components: Preparation and Fractionation,* 2d ed., G. D. Birnie Ed. Butterworths, London.

Fleischer, B., Fleischer, S., and Ozawa, H. (1969) Isolation and characterization of Golgi membranes from bovine liver. *J. Cell Biol.,* **43,** 53.

Hell, A., Rickwood, D. and Birnie, G. D. (1974) Buoyant density gradient centrifugation in solutions of metrizamide. In *Methodological Developments in Biochemistry,* Vol. 4, E. Reid Ed. Longmans Group Publishers, London.

Hinton, R. H., Norris, K. A., and Reid, E. (1970) Isolation of hepatic membrane fragments. In *Advances with Zonal Rotors,* E. Reid, Ed. Wolfson Bioanalytical Centre Publishers, London.

Hinton, R. H., Klucis, E., El-Aaser, A. A., Fitzsimmons, J. T. R., Alexander, P., and Reid, E. (1967) Zonal Centrifugation of hepatic microsomal material. *Biochem. J.* **105,** 14P.

Hu, A. S. L., Bock, R. M., and Halvorson, H. O. (1962) Separation of labeled from unlabeled proteins by equilibrium density gradient sedimentation. *Anal. Biochem.,* **4,** 489.

Leighton, F., Poole, B., Beaufay, H., Baudhuin, P., Coffey, J. W., Fowler, S., and de Duve, C. (1968) The large-scale separation of peroxisomes, mitochondria, and lysosomes from the livers of rats injected with Triton WR-1339. *J. Cell Biol.,* **37,** 482.

Meselson, M., Stahl, F. W., and Vinograd, J. (1957) Equilibrium sedimentation of macromolecules in density gradients. *Proc. Nat. Acad. Sci. (USA),* **43,** 581.

Nigam, V. N., Morais, R., and Karasaki, S. (1971) A simple method for the isolation of rat liver cell plasma membranes in isotonic sucrose. *Biochim. Biophys. Acta,* **249,** 33.

Pertoft, H., Philipson, L., Oxelfelt, P., and Hoglund, S. (1967) Gradient centrifugation of viruses in colloidal silica. *Virology,* **33,** 185.

Pertoft, H., and Laurent, T. C. (1969) The use of gradients of colloidal silica for the separation of cells and subcellular components. In *Modern Separation Methods of Macromolecules and Particles,* T. Gerritsen, Ed. Wiley-Interscience, New York.

Pertoft, H., Laurent, T. C., Laas, T., and Kagedal, L. (1978) Density gradients prepared from colloidal silica particles coated by polyvinylpyrrolidone (Percoll). *Anal. Biochem.,* **88,** 271.

Schumaker, V. N. (1967) Zone centrifugation. In *Advances in Bioligal and Medical Physics,* Vol. II, J. H. Lawrence and J. W. Gofman, Eds. Academic, New York, p. 245.

Sheeler, P. (1974) Reorienting density gradients and the SZ-14 rotor. In *Methodological Developments in Biochemistry,* Vol. 4, *Subcellular Studies,* E. Reid, Ed. Longmans Group Publishers, London.

Vedel, F., and D'Aoust, M. J. (1970) Rapid separation of ribosomal RNA by sucrose density gradient centrifugation with a fixed-angle rotor. *Anal. Biochem.,* **35,** 54.

Wilcox, H. G., Davis, D. C., and Heimberg, M. (1971) The isolation of lipoproteins from human plasma in zonal rotors. *J. Lipid. Res.* **12,** 160.

Additional Illustrative References to the Uses of *Rate* and *Isopycnic* Density-Gradient Centrifugation
Rate Separations

Beevers, L., and Poulson, R. (1972) Protein synthesis in cotyledons of *Pisum sativum L. Plant. Physiol.,* **49,** 476.

Birnie, G. D., Fox, S. M., and Harvey, D. R. (1972) Separation of polysomes, ribosomes and ribosomal subunits in zonal rotors. In *Subcellular Components; Preparation and Fractionation,* G. D. Birnie, Ed. Butterworths, London.

Boone, C. W., Harrell, G. S., and Bond, H. E. (1968) The resolution of mixtures of viable mammalian cells into homogeneous fractions by zonal centrifugation. *J. Cell Biol.,* **36,** 369.

Childress, W. J., Freedman, R. I., Koprowski, C., Doolittle, M. H., Sheeler, P., and Oppenheimer, S. B. (1979) Surface characteristics of separated subpopulations of mouse teratocarcinoma cells. *Exp. Cell Res.,* **122,** 39.

Evans, W. H. (1970) Fractionation of liver plasma membranes by zonal centrifugation. *Biochem. J.,* **116,** 833.

Johnston, I. R., Mathias, A. P., Pennington, F., and Ridge, D. (1968) The fractionation of nuclei from mammalian cells by zonal centrifugation. *Biochem. J.,* **109,** 127.

Kung, F. C., and Glaser, D. (1977) Synchronization of *Escherichia coli* by zonal centrifugation. *Appl. Environ. Micro.,* **34,** 328.

Lee, T., Swartzendruber, D. C., and Snyder, F. (1969) Zonal centrifugation of microsomes from rat liver: Resolution of rough and smooth-surfaced membranes. *Biochem. Biophys. Res. Commun.,* **36,** 748.

McCarty, K. S., Stafford, D., and Brown, O. (1967) Resolution and fractionation of macromolecules by isokinetic sucrose density gradient sedimentation. *Analyt. Biochem.,* **24,** 314.

Nissen-Meyer, J. Abro, A., and Eikhom, T. S. (1979) Isolation of intracisternal type-A particles and associated high-molecular-weight RNA after cell disruption by nitrogen cavitation. *Anal. Biochem.,* **97,** 85.

Perucho, M., Molgaard, H. V., Shevrack, A., Pataryas, T., and Ruiz-Carrillo, A. (1979) An improved method for the preparation of undegraded polysomes and active messenger RNA from immature chicken erythrocytes. *Anal. Biochem.,* **98,** 464.

Pickett, C. B., Cascarano, J., and Johnson, R. (1977) Oxidative phosphorylation in rat liver mitochondria isolated by rate zonal centrifugation: Examination of Ficoll gradients and subpopulations of mitochondria. *J. Bioenerg, Biomem.,* **9,** 271.

Rickard, K. A., Dunleavy, L., Brown, R., and Kronenberg, H. (1974) Fractionation of normal human bone marrow on albumin gradients. *Aust. J. Exp. Biol. Med. Sci.,* **52,** 767.

Rocha, V. and Ting, I. P. (1970) Preparation of cellular plant organelles from spinach leaves. *Arch. Biochem. Biophys.,* **140,** 398.

Storrie, B., and Attardi, G. (1973) Expression of the mitochondrial genome in HeLa cells. *J. Biol. Chem.,* **248,** 5826.

Swick, R. W., Strange, J. L., Nance, S. L., and Thomson, J. F. (1967) The heterogeneous distribution of mitochondrial enzymes in normal rat liver. *Biochemistry,* **6,** 737.

Wilson, M. A., and Cascarano, J. (1972) Biochemical heterogeneity of rat liver mitochondria separated by rate zonal centrifugation. *Biochem. J.* **129,** 209.

Isopycnic Separations

Barber, A. A., Harris, W. W., and Anderson, N. G. (1966) Isolation of native glycogen by combined rate-zonal and isopycnic centrifugation. *Nat. Cancer Inst. Monogr.,* **21,** 285.

Birnie, G. D., Rickwood, D., and Hell, A. (1973) Buoyant densities and hydration of nucleic acids, proteins and nucleoprotein complexes in metrizamide. *Biochim. Biophys. Acta* **331,** 283.

Bretz, U. (1973) Resolution of three distinct populations of nerve endings of rat brain by zonal isopycnic centrifugation. In *European Symposium of Zonal Centrifugation in Density Gradient,* G. Dorvyl, Ed. Editions Cite Nouvelle, Paris.

Glisin, V., Crkvenjakov, R., and Byus, C. (1974) Ribonucleic acid isolated by cesium chloride centrifugation. *Biochemistry,* **13,** 2633.

Hell, A., MacPhail, E., and Birnie, G. D. (1974) Buoyant density separation of nucleic acids in sodium iodide. In *Methodological Developments in Biochemistry,* Vol. 4, *Subcellular Studies,* E. Reid, Ed. Longmans Group Publishers, London.

Hopkins, H. A., Sitz, T. O., and Schmidt, R. R. (1970) Selection of synchronous *Chlorella* cells by centrifugation to equilibrium in aqueous Ficoll. *J. Cell Physiol.,* **76,** 231.

Johnson, C., Attridge, T., and Smith, H. (1973) Advantages of the fixed-angle rotor for the separation of density labelled from unlabelled proteins by isopycnic equilibrium centrifugation. *Biochim. Biophys. Acta,* **317,** 219.

Leif, R. C., and Vinograd, J. (1964) The distribution of buoyant density of human erythrocytes in bovine albumin solutions. *Proc. Nat. Acad. Sci. (USA),* **51,** 520.

Lewis, D. S., Cellucci, M. D., Masaro, E. J., and Yu, B. P. (1979) An improved method for the isolation of adipocyte plasma membranes. *Anal. Biochem.* **96,** 236.

Michel, J. M., and Michel-Wolwertz, M. R. (1970) Fractionation and photochemical activities of photosystems isolated from broken spinach chloroplasts by sucrose-density gradient centrifugation. *Photosynthetica,* **4,** 146.

Morre, D. J., Hamilton, R. L., Mollenhauer, H. H., Mahley, R. W., Cunningham, W. D., Cheetham, R. D., and Lequire, V. S. (1970) Isolation of a Golgi-rich fraction from rat liver. I. Method and morphology. *J. Cell. Biol.,* **44,** 484.

Morre, D. J., Yunghans, W. N., Vigil, E. L., and Keenan, T. W. (1974) Isolation of organelles and endomembrane components from rat liver. In *Methodological Developments in Biochemistry,* Vol. 4, *Subcellular Studies,* E. Reid, Ed. Longmans Group Publishers, London.

Perry, R. P., and Kelley, D. E. (1966) Buoyant densities of cytoplasmic ribonucleoprotein particles of mammalian cells: Distinctive character of ribosome subunits and rapidly labeled components. *J. Molec. Biol.* **16,** 255.

Pollak, J. K., and Woog, M. (1971) Changes in the properties of two mitochondrial populations during the development of embryonic chick liver. *Biochem. J.* **123,** 347.

Pollak, J. K., and Munn, E. A. (1970) The isolation by isopycnic density gradient centrifugation of two mitochondrial populations from livers of embryonic and fed and starved rats. *Biochem. J.,* **117,** 913.

Price, C. A. (1973) Separation of chloroplasts by isopycnic, rate zonal and continuous flow density gradient centrifugation. In *European Symposium of Zonal Centrifugation in Density Gradient,* G. Dorvyl, Ed. Editions Cite Nouvelle, Paris.

Prospero, T. D., and Hinton, R. H. (1972) Isolation of plasma membrane fragments from hepatomas. In *Methodological Developments in Biochemistry,* Vol. 3, *Advances with Zonal Rotors,* E. Reid, Ed. Longmans Group Publishers, London.

Schnaitman, C. A., Erwin, V. G., and Greenawalt, J. P. (1967) The submitochondrial localization of monoamine oxidase. *J. Cell Biol.,* **32,** 719.

Sheeler, P., Moore, J., Cantor, M., and Granik, R. (1968) The stored polysaccharide of *Polytomella agilis. Life Sci.* **7,** 1045.

Tolbert, N. E. (1974) Isolation of subcellular organelles of metabolism on isopycnic sucrose gradients. In *Methods in Enzymology,* Vol. 31 (S. Fleischer and L. Packer, Eds.) Academic, New York.

Vinograd, J., and Bruner, R. (1966) Band centrifugation of macromolecules in self-generating density gradients. III. Conditions for convection-free band sedimentation. *Biopolymers,* **4,** 157.

Wattiaux, R., and Wattiaux-De Coninck, S. (1970) Distribution of mitochondrial enzymes after isopycnic centrifugation of a rat liver mitochondrial fraction in a sucrose gradient: Influence of the speed of centrifugation. *Biochem. Biophys. Res. Commun.,* **40,** 1185.

Werner, S., and Neupert, W. (1972) Functional and biogenetical heterogeneity of the inner membrane of rat-liver mitochondria. *Eur. J. Biochem.,* **25,** 379.

Williamson, R. (1969) Purification of DNA by isopycnic banding in CsCl in a zonal rotor. *Anal. Biochem.,* **32,** 158.

SIX

Density Gradient Centrifugation in Zonal Rotors

In most laboratories, density gradient centrifugation is carried out by using tubes in either swinging-bucket or fixed-angle rotors. Among the limitations imposed by this approach are (1) the restricted capacity of the rotors, especially those that are capable of very high speeds (see Appendixes) and (2) the absence of sector shape in the centrifuge tubes. *Zonal rotors* offer greatly increased sample and gradient capacity and the advantages of particle sedimentation in sector-shaped chambers. The typical zonal rotor is a hollow, cylindrical bowl subdivided into a number of sector-shaped compartments by vertical *septa* (or *vanes*) that radiate from the central *core* to the bowl wall. Each chamber is analogous to an individual centrifuge tube. All chambers are simultaneously filled with a single density gradient that is used to fractionate a single sample.

Zonal rotors can be subdivided into two major categories: (1) *rotating-seal* or *dynamically unloaded* zonal rotors and (2) *reorienting gradient (reograd)* or *statically unloaded* zonal rotors. The rotating-seal zonal rotor is loaded with gradient and sample *while the rotor is spinning;* following separation of the particles in the density gradient, the gradient and the entrained particle zones are displaced from the rotor, also while it is spinning. Dynamic loading and unloading are achieved by using a special seal assembly that is comprised of rotating and static faces. Channels through both halves of the seal provide continuous communication with the interior of the spinning rotor. In contrast, reorienting gradient zonal rotors are unloaded at rest (i.e., statically), although certain models can be loaded dynamically. Static unloading requires that the density gradient in each of the rotor chambers reorient from the radial to the vertical position as the rotor is slowly brought to rest. The phenomenon is similar to that described in Chapter 5 in connection with the use of fixed-angle rotors for density gradient separations.

DEVELOPMENT OF ZONAL ROTORS

The origins of both rotating-seal and reorienting gradient zonal rotors can be traced to Norman G. Anderson, whose contributions to centrifugal technology and whose innovations in particle and molecular separations are now almost legendary. Although some preliminary work had been done earlier, the development of zonal rotors began in earnest in 1961 at the Biology Division of the Oak Ridge National Laboratory under the direction of Anderson. The work there was supported jointly by the National Institutes of Health and the Atomic Energy Commission and ultimately became known as the "MAN" (an acronym for "molecular anatomy") program; the program remained active until 1977. During these years, more than 50 different rotors were developed, and these served as the forerunners of most of the zonal and continuous-flow rotors that are presently available commercially. At the outset, Anderson subdivided the zonal rotor development program into two major categories: (1) low-speed or *A-type* rotors, and (2) higher-speed or *B-type* rotors. In each series, successive rotors were identified by numbers (e.g., A-I, A-II, A-III, etc. and B-I, B-II, B-III, etc.) Because they represented either major technological or innovative advances or were the forerunners of successful commercial counterparts, a number of these rotors are briefly described (see Table 6-1).

ROTATING-SEAL (DYNAMICALLY UNLOADED) ZONAL ROTORS

Anderson's first rotor, the A-I, was a conventional, low-speed swinging-bucket type, but the buckets accepted specially constructed sector-shaped glass centrifuge tubes (Anderson, 1955). The major contribution of the studies made using this rotor was demonstrating the effectiveness with which density gradient centrifugation could be carried out when the sedimenting particles had an uninterrupted path through the gradient (i.e., no wall effects). Anderson's A-II rotor (Fig. 6-1) was a modified swinging-bucket rotor in which the tubes could be loaded while the rotor was spinning (Albright and Anderson, 1958). Each tube was fitted with a central, stainless steel cannula, one end of which opened at the bottom of the tube

Table 6-1 Major Stages in the Development of Zonal Rotors

Year	Original Designation	Commercial Equivalent	Developed by	Comments	Reference
			Rotating-Seal Zonal Rotors		
1955	A-I	—	N. G. Anderson	Swinging-bucket rotor using special sector-shaped tubes	Anderson, 1955
1958	A-II	—	N. G. Anderson	Dynamically-loaded tubes in a swinging-bucket rotor; tubes unloaded statically	Albright and Anderson, 1958
1956–1962	A-III	—	N. G. Anderson	First zonal rotor; first rotor loaded and unloaded dynamically; no rotating seals	Anderson, 1956
1962	A-IV	—	N. G. Anderson	First zonal rotor loaded and unloaded dynamically through a rotating-seal assembly	Anderson, 1962a
1962	B-II	—	N. G. Anderson	First high-speed zonal rotor, loaded and unloaded dynamically; immediate forerunner of first commercial zonal rotor	Anderson, 1962a Anderson, 1962b
1964	B-IV	B-IV[a]	N. G. Anderson	First commercial zonal rotor	Anderson et al., 1964a
1964	A-XII	A-XII[b] A-type[c]	N. G. Anderson	First commercial low-speed zonal rotor; transparent end caps for continuous visual monitoring	Anderson et al., 1966

Year	Rotor	Rotor	Developer	Description	Reference
1965	B-XIV and B-XV	B-XIV[b-d], B-XV[b-d]	N. G. Anderson	First zonal rotors with *removable* seal assembly	Anderson, 1966
1967	B-XXIII	—	N. G. Anderson	First zonal with removable seal and choice of core or wall loading and unloading	Anderson et al., 1968
1968	B-XXIX	B-XXIX[3,4] and B-XXX[3,4]	N. G. Anderson	Removable seal and choice of core or wall loading and unloading through wall taper plus cloverleaf groove	Anderson et al., 1969

Reorienting-Gradient Zonal Rotors

Year	Rotor	Rotor	Developer	Description	Reference
1964	A-VII	—	N. G. Anderson	First reorienting gradient zonal rotor	Anderson et al., 1964b
1969	A-XVI	—	N. G. Anderson		Elrod et al., 1969
1969	RG-1	—	P. Sheeler and J. R. Wells	First zonal rotor developed outside MAN Program	Sheeler and Wells, 1969
1970-1971	SZ-14 and SZ-20	SZ-14[e] and TZ-28[e]	P. Sheeler and J. R. Wells	First dynamically loaded but statically unloaded zonal rotor; first reograd zonal to become available commercially	Sheeler et al., 1971

[a] Made by Spinco Division of Beckman Instruments, Inc. No longer in manufacture.
[b] Made by Damon Industries—I.E.C. No longer in manufacture.
[c] Made by and available from Measuring and Scientific Equipment, Inc. (United Kingdom).
[d] Made by and available from Spinco Division, Beckman Instruments, Inc.
[e] Made by and available from Sorvall Division, E. I. Dupont and Company, Inc.

Fig. 6-1 The A-II rotor. (Courtesy of Dr. N. G. Anderson.)

while the other end passed up through the tube cap and flared to form
a receptacle. When the tubes were swung into the horizontal position by
acceleration of the rotor, the receptacles engaged a series of radially directed spouts emerging from an axial distributor that was mounted on and
spinning with the rotor. The sample and the density gradient were fed
into the spinning distributor, where centrifugal force carried the stream
through each of the spouts and into the tube cannulas. In this way the
sample and the supporting density gradient gradually and simultaneously
filled each of the tubes. Gradients and particle zones were recovered from
the tubes after the rotor was decelerated to rest.

In the A-III rotor (Fig. 6-2), centrifuge tubes were eliminated altogether, and the A-III may be considered the first "true" zonal rotor

(Anderson, 1956). In this rotor, dynamic loading *and* unloading were possible, and each of the rotor's chambers had sector shape. The density gradient was introduced to the wall of the spinning rotor light end first through tubes leading radially from an annular chamber located just beneath the upper end cap. The sample was then injected through an axial opening in the rotor lid and onto the surface of the gradient (see Fig. 6-2). After particle separation had taken place, the gradient was displaced toward the center of the rotor by allowing dense solution to flow to the rotor wall. The centripetally displaced gradient, together with the separated particle zones, exited the rotor through drain tubes inserted into the lower end cap adjacent to the core. It is to be noted that in the A-III rotor, both dynamic loading and unloading were achieved without the use of a rotating seal assembly. The immediate successor to the A-III (i.e., the A-IV) demonstrated the feasibility of loading and unloading the rotor through rotating seals, thereby providing the closed system necessary for working with pathogenic or otherwise hazardous materials (Anderson, 1962a).

The B-II rotor (Anderson, 1962a, 1962b) was the first zonal rotor in

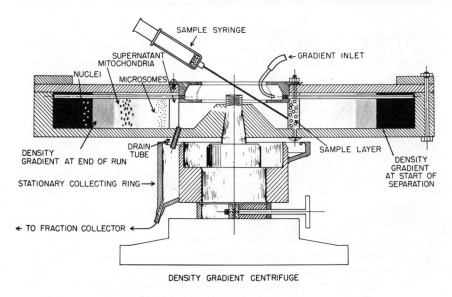

DENSITY GRADIENT CENTRIFUGE

Fig. 6-2 The A-III rotor. (Courtesy of Dr. N. G. Anderson.)

which extensive biological separations were attempted (Fig. 6-3). The rotor chamber was divided into 36 sector-shaped compartments by thin septa inserted into slots in the core. The core itself was doubly constricted in order to channel the gradient and separated zones in the lower and upper halves of the rotor toward the core openings during the unloading operation (see below). Communication with the rotor wall was achieved through slots in the upper end cap and the top of the core. Heat of friction produced between the static upper half of the seal assembly and the rotating lower half was reduced by circulating cold water through channels in the static seal.

Although it is no longer in production, Anderson's B-IV rotor (Fig. 6-4) was the first zonal rotor to be produced commercially (by the Spinco Division of Beckman Instruments). In the B-IV, the number of septa was reduced to four, and these contained the channels that lead to the rotor wall. The core surface was flattened and gently tapered to a single, central exit port in each of the rotor's chambers. The B-IV rotor could be used at speeds of up to 40,000 rpm in a special version of the model L ultracentrifuge (called the model ZU) (Anderson et al., 1964a).

The A-IX rotor was the immediate forerunner of the first commercial low-speed zonal rotor, the A-XII, produced by International Equipment Corporation (Anderson et al., 1966). The A-IX employed eight septa and the A-XII only four, but in most other respects the two rotors were the same (Fig. 6-5). Both rotors had transparent (Lucite) upper and lower end plates secured by aluminum retaining rings. The separation of particle zones in these rotors could be followed visually using a light source mounted in the centrifuge chamber below the rotor (Fig. 6-6). The A-XII rotor is still in production and available from Measuring and Scientific Equipment Company (United Kingdom). It can be operated at speeds of up to 5000 rpm.

Removable Seals   The rotating seal that characterizes all dynamically unloaded zonal rotors consists of two principal parts: (1) a static upper half usually constructed of stainless steel and (2) the rotating lower half, usually molded from Rulon or graphite (see Fig. 6-7). The apposing faces of the seal are pressed together by spring tension. Two independent sets of channels or "lines" pass through the seal. The *center* lines are axially positioned and in continuous communication during operation. In most zonal rotors these lines lead to the surface of the core. The two seal halves also contain an annular array of channels at the same radius. These become congruent with a certain frequency as the seal faces rotate relative to one another. During the congruent intervals, fluid can pass through the two halves of the seal. The annular lines pass through the core and into the septa, where they then radiate to the rotor wall.

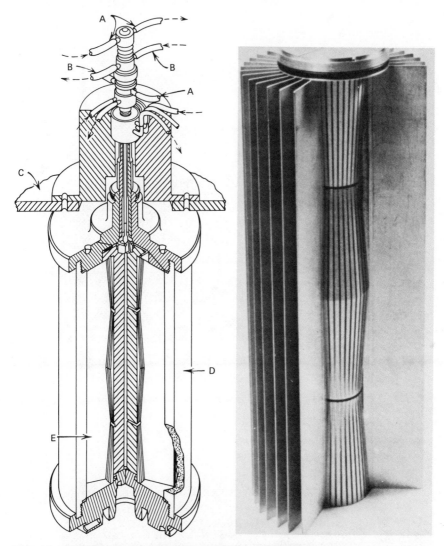

Fig. 6-3 The B-II rotor. *Left*, cutaway section view through the rotor and the centrifuge chamber cover. *Right*, doubly constricted core with septa. *A*, fluid lines to rotor center; *B*, fluid lines to rotor wall; *C*, top of centrifuge chamber; *D*, rotor wall; *E*, septum.

Fig. 6-4 The B-IV rotor. *Left*, assembled; *right*, disassembled; *a*, top end cap; *b*, feed lines and seal; *c*, bearing system; *d*, bottom end cap; *e*, rotor wall; *f*, core with four septa.

Fig. 6-5 The A-XII rotor. (Top, assembled; bottom, disassembled.)

From the outset, one of the principal problems with dynamically loaded and unloaded zonal rotors was the amount of wear that the seal faces incurred during the course of a run. Since the seal assembly was used only during loading and unloading (and not during the separation of particles at speed), much of the wear was unnecessary. A major step forward in zonal rotor development was the introduction of the removable-seal assembly. This change provided for the detachment of the seal

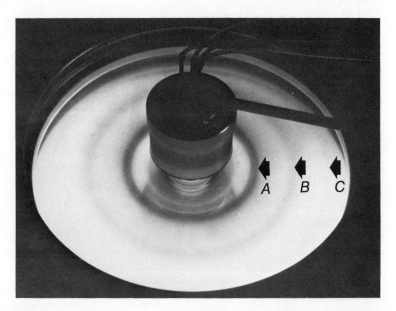

Fig. 6-6 In the A-XII rotor the separation of the components of the sample can be followed visually. This photograph was taken during the course of fractionation of a liver homogenate. Marker analysis of the three zones that can be discerned indicated that these contained microsomes and cytosol (zone *A*), mitochondria (zone *B*), and cell nuclei and plasma membrane fragments.

after sample and gradient loading and its reattachment prior to the unloading operation. Consequently, during the course of a run (which might be lengthy and carried out at a much higher speed than that used during the loading–unloading operations), there was no seal wear. The first commercial versions of the removable-seal-type zonal rotors were the B-XIV and the B-XV (Fig. 6-7) (Anderson, 1966). These are still produced by several companies and can be operated in conventional ultracentrifuges. In the early versions, the removable element contained the static half of the seal and the fixed element contained the rotating half. In current models produced by some companies, the removable element contains both halves of the seal.

Edge-Unloading Zonal Rotors   Up through the B-XV model, the particle mixture to be fractionated and the separated zones produced during a run respectively entered the spinning rotor and were later collected through channels that lead to the surface of the core. The core was especially tapered to maximize the efficiency of these movements (see below). An-

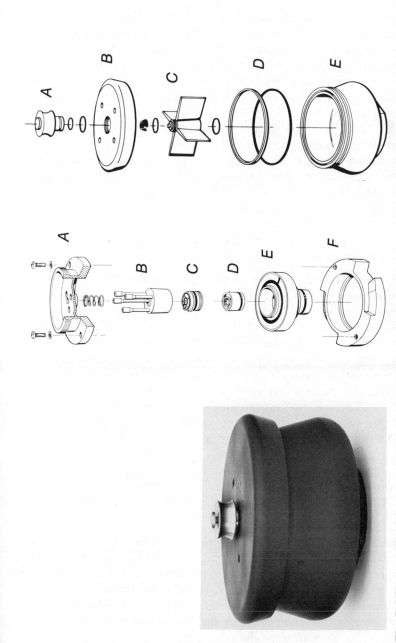

Fig. 6-7 The B-XV rotor (Beckman's model Ti-15). *Left*, assembled rotor. *Center*, exploded view of rotating seal assembly (*A*, bearing cover; *B*, manifold; *C*, stationary portion of the coaxial seal assembly; *D*, rotating portion of the seal assembly; *E*, bearing; *F*, bearing housing). *Right*, exploded view of the rotor (*A*, cap; *B*, lid; *C*, core and septa; *D*, extrusion ring and gasket; *E*, rotor bowl). (Courtesy of Beckman Instruments, Inc.)

other major advance in zonal rotor design was the addition of a tapered surface at the rotor edge that made it possible to efficiently introduce or collect particle zones from the wall of the rotor as well as from the core. The first successful edge-unloaded zonal rotor was the B-XXIII Oak Ridge prototype (Anderson et al., 1968). This lead to a modified version called the B-XXIX (Fig. 6-8), which presently has several commercial counterparts. In essence, the B-XXIX (and the B-XXX) zonal rotor is identical to the B-XV (and B-XIV) with the exception of the tapered wall and modified septa that permit the operator to select either core or edge unloading (Anderson et al., 1969a). Some companies manufacture separate B-XV and B-XXIX rotors, whereas others offer inserts that convert one type to the other (Fig. 6-9).

Operation of a Rotating-Seal Zonal Rotor   Since the B-XXIX rotor has features found in nearly all dynamically unloaded zonal rotors, a description of its operation may be considered representative.

Loading the Density Gradient   The empty, assembled rotor is placed in the centrifuge and the seal assembly, mounting hardware, and all connecting lines are attached. The rotor is accelerated to the loading speed (about 2500 rpm) and the density gradient pumped light end first from the gradient maker into the edge line opening of the static seal. The gradient flows down into the rotating half of the seal and then through channels in each of the septa to the rotor wall. Fluid entering the rotor bowl at the exit ports of each septum is displaced centripetally guided by both a cloverleaf equatorial groove (a "super circle" of four equal maximum radii and four equal minimum radii) and the vertically tapered wall surfaces (Fig. 6-10A). A cushion is usually added to the dense end of the gradient until a small quantity of the light end emerges from the rotor through the core lines (Fig. 6-10B). At this point the rotor bowl is full.

Some workers prefer to fill the rotor with water prior to the loading operation. During loading, the water is displaced from the rotor as it is replaced by the density gradient and the cushion.

Loading the Sample   The direction of flow through the rotor is reversed as the sample is pumped into the core lines (Fig. 6-10C). To ensure that the sample zone formed inside the rotor is uniformly thick, a light overlay solution is pumped into the core lines, thereby displacing the sample radially beyond the tapered region of the core (Fig. 6-10D). Addition of the sample and overlay is accompanied by the displacement of some of the cushion out of the rotor through the edge lines.

Separation of Particles   With the cushion, the gradient, the sample, and the overlay loaded, the seal assembly is lifted off the spinning rotor. The

Fig. 6-8 The B-XXIX rotor (top, assembled; bottom, disassembled). Unlike the B-XV, this zonal rotor can be unloaded through either the core region or the edge of the rotor bowl (see also Fig. 6-10).

113

Fig. 6-9 Bowl insert and replacement core–septa piece that converts the B-XV to a B-XXIX rotor. (Courtesy of Beckman Instruments, Inc.).

channels leading to the core surface and to the bowl wall are closed off by the attachment of a vacuum sealing cap. The rotor is then accelerated to the operating speed (and the centrifuge vacuum turned on). Particles in the sample sediment radially to form a number of concentric cylindrical zones (Fig. 6-10E).

Unloading the Gradient   Once particle separation is completed, the rotor is decelerated to the unloading speed (also about 2500 rpm), the vacuum cap removed, and the seal assembly reattached. The density gradient and the entrained particle zones can be collected from the rotor through either the core lines or the edge lines. To collect through the core lines, cushion is pumped to the edge of the rotor until the entire gradient has been displaced (Fig. 6-10F). Using this procedure, the separated particles are collected in order of increasing sedimentation coefficient or banding density.

To collect from the edge of the rotor bowl, additional overlay is pumped through the core lines. In this approach, it is the dense end of the gradient that is collected first (Fig. 6-10G). Usually, the density gradient and the separated particles are collected as a continuous sequence of fractions that are subjected to additional experimental analysis.

Edge-loading and edge-unloading zonal rotors offer a number of advantages over their core-unloading counterparts. First, they can be unloaded by using water as the displacing medium. This is considerably less costly and also less time-consuming than core unloading by using dense salt solutions or dense, viscous solutions of sucrose. Edge-unloading zonal rotors are also effective for flotation experiments in which the sample is introduced below the dense end of the density gradient (i.e., at the rotor wall) and the particles floated to their isopycnic positions. In some instances it is desirable to remove some of the particles that have sedimented into the dense end of the gradient and to replace the gradient with fresh, particle-free medium. This is possible only with the edge-unloading versions or adaptations of zonal rotors.

Other Dynamically Unloaded Zonal Rotors   Rotating-seal zonal rotors are manufactured by a number of different instrument companies (see the Appendixes). The most popular versions and those in most general use are adaptations of the original A and B series models developed by N. G. Anderson at the Biology Division of the Oak Ridge National Laboratory. Most of the important specifications (configuration, size, capacity, operation, etc.) set down during the development of these rotors have been adhered to in the production of the commercial versions. There are, however, a number of dynamically unloaded zonal rotors that are not direct counterparts of Anderson's A and B series. The more popular or commonly encountered among these are the JCF-Z (Beckman Instruments), the HS (Measuring and Scientific Equipment), and the Z-15 (Damon/International Equipment Corp.). Model JCF-Z is the only zonal rotor produced by the Spinco division of Beckman Instruments that may be used with their superspeed line of centrifuges (e.g., model J2-21). It is similar to the B-XV with somewhat larger capacity and a modified and simplified rotating seal system. Of special interest are the interchangeable cores that permit this rotor also to be used for reorienting gradient (see below) and continuous-flow (Chapter 10) work.

The HS and Z-15 rotors are quite similar, as both are scaled-down versions of the original A-XII with Lucite upper and lower end caps. However, the HS and the Z-15 may be used at higher speeds than the original A-XII (i.e., ≤10,000 rpm).

K-Series Zonal Rotors   During the last several years of the MAN program at Oak Ridge, a number of zonal rotors were developed that had especially high capacities and some unusual features. These rotors are known generally as the K series (Reimer et al., 1967; Anderson et al., 1969b). Interchangeable cores permit their use for both zonal density gradient separations and continuous-flow harvesting of small particles (especially viruses). In these rotors, there were two axially positioned rotating seals, one mounted at the top of the rotor and the other at the bottom. Liquid entered the spinning rotor through one seal system and exited through the other. Commercial versions of the K-series zonal rotors are used

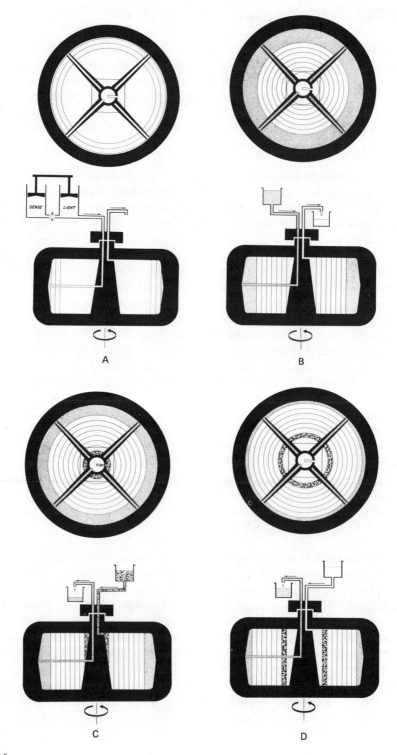

A

B

C

D

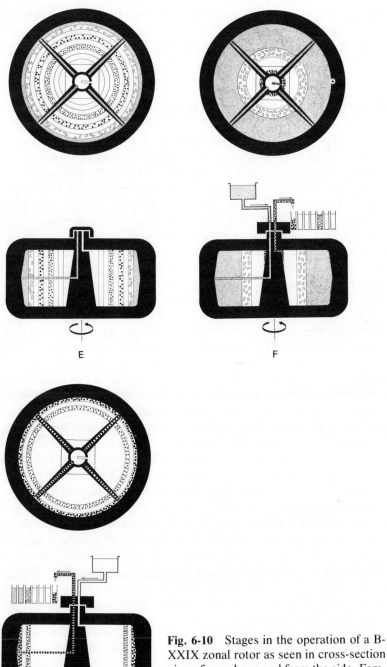

Fig. 6-10 Stages in the operation of a B-XXIX zonal rotor as seen in cross-section views from above and from the side. Families of lines within the rotor bowl represent planes of equal density within the gradient (see text for description of each stage).

E

F

G

117

primarily by pharmaceutical companies for quantitative preparations of vaccines.

REORIENTING GRADIENT (REOGRAD) ZONAL ROTORS

The origins of reorienting gradient zonal rotors can also be traced to the work of N. G. Anderson. Two principal features distinguish these rotors from dynamically unloaded zonals: (1) the density gradient is unloaded with the rotor at rest (i.e., static); and (2) no rotating seal assembly is required. Depending on the model used, gradient and sample may be *loaded* either statically or dynamically. Static loading of the rotor is followed by a smooth and gradual transition in which the vertical gradient is transformed into one that is radially distributed within the rotor bowl; that is what is meant by "gradient reorientation." In versions of this rotor in which dynamic loading is possible, the radial density gradient is established immediately and in much the same fashion as in a rotating-seal zonal rotor. Unloading is carried out after the rotor is brought to rest. During deceleration, the density gradient (and separated particles) is reoriented from the radial to the vertical position.

Gradient Reorientation Before we consider the development of reograd zonal rotors and the operation of a representative model, it is of value to consider the phenomenon of gradient reorientation as illustrated in Fig. 6-11. For simplicity, the zonal rotor is depicted as a hollow cylinder with horizontal lid and base. With the rotor at rest, the chamber is filled with a cushion, density gradient, sample, and overlay (Fig. 6-11A). Isodense surfaces in the gradient are depicted as horizontal lines. Prior to acceleration of the rotor, the only force acting on the isodense surfaces is *gravity* (i.e., F_g). As the rotor chamber is slowly accelerated, each isodense surface experiences an additional force, namely, *centrifugal force* (i.e., F_C). At any specific rotational speed, the ratio $F_C:F_g$ across an isodense surface increases in proportion to the distance from the axis of rotation. The result is that successive isodense surfaces are bowed downward at the axis and upward at the periphery so that they are transformed into a family of *paraboloids of revolution* with their foci lying on the axis of rotation (Fig. 6-11B).

The vertical height Z above the focus for any point on the paraboloidal surface is given by the relationship

$$Z = \frac{\omega^2 x^2}{2g} \qquad (6\text{-}1)$$

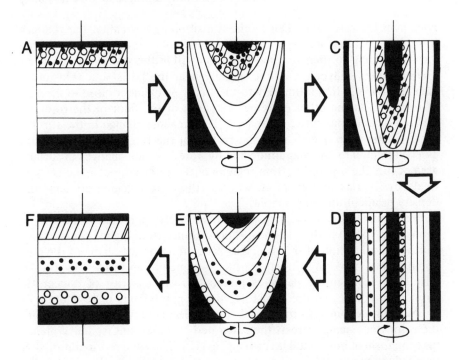

Fig. 6-11 Gradient reorientation and particle separation in an idealized zonal rotor. Families of lines (horizontal, vertical, parabolas) represent planes of equal density. Dark zones above sample and below gradient in stage A are overlay and cushion. See text for descriptions of each stage.

where ω is the angular velocity, x is the radial distance to the point on the paraboloid, and g is the earth's gravitational constant. From this relationship it can be seen that the paraboloids become increasingly steep as the rotor accelerates (Fig. 6-11C). At high speed, where the $F_C:F_g$ ratio is very high, the isodense surfaces approach "verticality." That is, the foci of the paraboloids lie great distances below the bottom of the rotor, so that within the body of the rotor itself, the family of paraboloidal surfaces may be considered to be a family of concentric cylinders (Fig. 6-11D, right). In most zonal rotors, verticality is achieved by the time that the rotor reaches or exceeds 1000 rpm.

As a result of reorientation, the original vertical density gradient has been replaced by a radial one. The densest region of the gradient, which initially occupied the lower region of the rotor chamber, forms a cylindrical zone near the rotor wall; the lightest region of the gradient, which initially resided near the top of the rotor chamber, forms a cylindrical zone near the rotor axis. During the transition to verticality, the changes in surface area that are experienced by equal volumes of gradient material vary according to their original locations within the stationary rotor. At one extreme, a layer or a zone initially near the lid is squeezed into a small paraboloid that ultimately forms a narrow cylinder close to the axis of rotation. At the other extreme, a zone near the floor of the rotor ultimately covers the entire rotor wall. The smallest change in the surface area:volume ratio occurs near the middle of the rotor. Since resolution can be affected by the magnitude and the rate at which changes in surface area occur, the top and bottom regions of the rotor are occupied respectively by overlay and cushion, whereas the central region contains the density gradient and the sample.

Returning to Fig. 6-11, with verticality closely approached, the sample forms a narrow cylindrical zone a short distance from the axis of rotation. Further acceleration of the rotor has little effect on the shape of the paraboloids but markedly affects particle sedimentation rates (Fig. 6-11D, left). The sedimenting particles gradually form a series of concentric cylindrical zones that are transformed into paraboloids of revolution of diminishing steepness as the rotor is slowly decelerated (Fig. 6-11E). With the rotor once again at rest, the separated particles and isodense planes take the form of horizontal layers (Fig. 6-11F). The zone reorientation that takes place in a zonal rotor during acceleration is vividly depicted in the series of photographs in Fig. 6-12. These photographs were obtained by using a specially fabricated transparent zonal rotor in the author's laboratory.

Early Reograd Zonal Rotors   The first reograd zonal rotor was built by Anderson in 1964 and was designated rotor A-VII (Anderson et al., 1964). The rotor consisted of a small methacrylate cylinder divided into four compartments by radial septa. The floor of the rotor was tapered downward to form an axial vertex for introduction and collection of materials; this was achieved statically by using a pipette inserted through an opening in the lid. The rotor was mounted in a brass holder attached to a heavy flywheel. The inertia produced by this large total mass provided the smooth and extended acceleration and deceleration found necessary for gentle and efficient gradient reorientation. In a more sophisticated model, the A-XVI, built by Anderson in 1969, the loading and unloading lines were incorporated into a narrow core piece and both the lid and floor were tapered to form funnels. One of the core lines descended to the vertex of the conical chamber floor, and the other line opened near the vertex of the lid. As with the A-VII, gradient and sample were loaded and

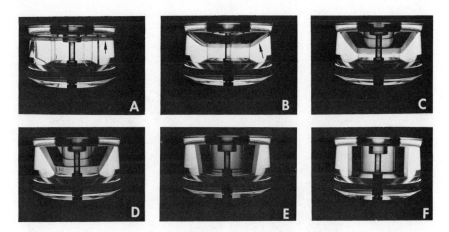

Fig. 6-12 Reorientation of an isodense plane (arrow) during the acceleration of a zonal rotor: *A*, rotor at rest (0 rpm); *B* 100 rpm; *C*, 180 rpm; *D*, 250 rpm; *E*, 400 rpm; *F*, 1,000 rpm. Increasingly steep paraboloids of revolution are formed until verticality is attained. These photographs were taken by using a transparent (Lucite) version of the TZ-28 rotor described in the text.

unloaded statically, and gradual and smooth reorientation was achieved by attaching the rotor to a heavy flywheel (Elrod et al., 1969).

The SZ-14 and TZ-28 Reograd Zonal Rotors The first zonal rotors designed and constructed outside the MAN program were the reorienting gradient rotors of the author (Sheeler and Wells, 1969; Sheeler et al., 1971). Probably the most significant single departure in design in the author's rotor was the incorporation of an annular V-shaped groove in the rotor floor. It is from this groove that the density gradient and the separated zones of particles are collected at the end of centrifugation. At the same time, the groove frees the axial region of the rotor so that it may be lowered down and around the drive shaft of the centrifuge. No flywheel is used.

In the first of these reograd rotors, built in 1969, the chamber was divided into 12 compartments by septa and static loading and unloading achieved through 12 steel tubes that descended to the vertex of the V groove (Fig. 6-13). Dynamic gradient and sample loading was also possible and was carried out by allowing gradient and sample to be successively carried by centrifugal force across an outwardly tapered opening in the rotor lid and onto the surface of each of the septa. In the absence of a flywheel, gradual rotor acceleration and deceleration was achieved by

Fig. 6-13 Early experimental reorienting gradient zonal rotor. *Top*, disassembled; *bottom*, rotor assembled using the transparent lid.

using an attachment to the centrifuge (an electronic module called a *rate controller*) that could be switched into the centrifuge circuitry in place of the regular speed control. The rate controller fed a small and variable amount of direct current to the centrifuge motor so that the rotor could be slowly accelerated. During the deceleration phase, the current fed to the centrifuge motor served to prolong the deceleration time. Above 1000 rpm, the speed of the rotor was controlled by the conventional circuitry of the centrifuge.

After testing and evaluating several additional experimental versions, a final rotor design was adopted in 1970 (Sheeler, 1971). Gradient and sample loading was achieved through an axially mounted distributor and could be carried out either statically or dynamically, whereas unloading (carried out statically only) was affected through channels in the six radial septa that descended into the annular V groove in the rotor floor. The first commercial version of this rotor, called the SZ-14, was manufactured by DuPont-Sorvall Instruments in 1971 and was used in all of their superspeed centrifuges (Sheeler and Wells, 1973; Sheeler, 1974). Rotor acceleration and deceleration during the reorientation period were controlled in the manner described above. The SZ-14 rotor has since been replaced by a high-speed titanium version designated model TZ-28 and can be used in the DuPont oil-turbine-drive ultracentrifuges as well as in their line of superspeed centrifuges. In the oil-turbine system, slow acceleration of the reograd rotor is obtained through a control module that varies the rate of oil flow across the small turbine at the base of the centrifuge drive unit. Slow deceleration is obtained by programming the oil pump to shut off once the rotor speed decreased below 1000 rpm. In the absence of the frictional drag that characterizes conventional direct, gear, or belt drives, the rotor's deceleration is markedly smoothed and prolonged.

Design and Operation of the TZ-28 Rotor   The TZ-28 rotor is shown in Figs. 6-14 and 6-15. The rotor chamber surrounds an axial shaft, and the wall of the chamber and the shaft are tapered at the bottom to form an annular V groove in the rotor floor. The unitized core–septa piece slides over the shaft and divides the chamber into six sector-shaped compartments. Two sets of channels called the *septa lines* and the *core lines* pass through the core–septa piece. The six septa lines descend at an angle from the surface of the core through the septa to open at the vertex of the annular V groove, whereas the six core lines open at the outer core wall, adjacent to each septum. Removably mounted on top of the core is a circular distributor. Two different distributors are used with the TZ-28; one of these is attached when the rotor is to be operated in the evac-

Fig. 6-14 The TZ-28 rotor. *Left*, disassembled rotor; *top right*, assembled with superspeed distributor; *bottom right*, assembled with ultraspeed distributor.

uated chamber of an ultracentrifuge and the other when the rotor is used in a superspeed centrifuge. The distributors contain two sets of radial channels that communicate with corresponding channels in the core–septa piece.

In the superspeed distributor, the core lines open into the centrifugal edge of an annular loading ring, whereas in the ultraspeed distributor the core lines open individually at the surface. In both distributors the septa lines converge on an axial chamber. The stempiece fits into the axial distributor chamber during the unloading operation and during static loading. A sealing cover is mounted over the ultraspeed distributor and seals the rotor during operation in the evacuated chamber of an ultracentrifuge. Sealed operation of the rotor may also be carried out in a superspeed centrifuge but is not mandatory.

Unsealed Operation (Fig. 6-16)   The empty rotor is placed in the centrifuge and the superspeed distributor attached. For *static* loading of the density gradient (Fig. 6-16A), tubing from the gradient maker is connected to the stempiece, which is then inserted into the axial opening of the distributor; the density gradient is pumped into the rotor beginning with the light end. The gradient enters the distributor and passes through the lines of the core–septa piece to the vertex of the annular V groove. As it emerges from the small openings at the vertex, it is displaced upward by the denser fluid flowing immediately behind. In this manner, the rotor is slowly filled with a gradient whose most dense region is at the bottom of the annular V groove and that becomes progressively lighter on rising vertically.

The stempiece is removed and the rotor slowly accelerated. Acceleration of the rotor causes the density gradient to undergo reorientation (Fig. 6-16B) until the radial gradient is established. Reorientation is virtually complete by the time the rotor attains 800 rpm, so that acceleration above this point can be carried out more quickly. Slow acceleration is achieved by using the rate controller module of the centrifuge.

The sample to be fractionated is introduced onto the reoriented gradient at about 1000 to 2500 rpm. This is accomplished by depositing the

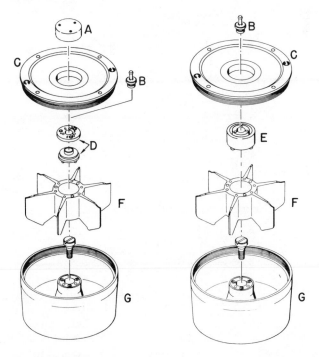

Fig. 6-15 Exploded views of the ultraspeed (left) and superspeed (right) TZ-28 rotor: *A*, sealing cover; *B*, stem piece; *C*, lid; *D*, ultraspeed distributor; *E*, superspeed distributor; *F*, core/septa piece; *G*, rotor bowl.

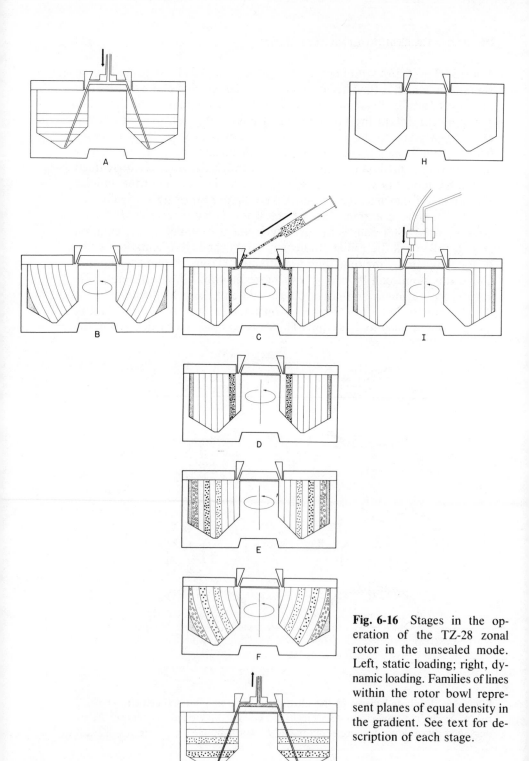

Fig. 6-16 Stages in the operation of the TZ-28 zonal rotor in the unsealed mode. Left, static loading; right, dynamic loading. Families of lines within the rotor bowl represent planes of equal density in the gradient. See text for description of each stage.

sample into the loading ring of the distributor (Fig. 6-16C). Centrifugal force sweeps the sample radially through the core channels of the distributor and core–septa piece and deposits it on the density gradient (Fig. 6-16D). With the sample loaded, the rotor may now be accelerated (or decelerated) to a higher (or lower) operating speed. Separation of the particles in the sample zone into a series of concentric cylindrical zones is depicted in Fig. 6-16E. The annular V groove has asymmetrically sloping surfaces so that particles in that portion of the sample zone that initially descends into the centripetal region of the groove do not encounter the upwardly sloping centrifugal surface during sedimentation.

Following centrifugation at operating speed for the desired period of time, the rotor is decelerated. During deceleration, the density gradient and the entrained particle zones undergo reorientation (Fig. 6-16F). With the rotor once again at rest, the vertical density gradient is reestablished, and the separated particle zones form a series of horizontal layers in the gradient (Fig. 6-16G). The stempiece is reinserted into the distributor and the density gradient pumped (i.e., withdrawn) from the rotor. The gradient exits the rotor beginning with that portion occupying the annular V groove. Thus the separated particles are collected in order of decreasing sedimentation rate or decreasing density.

The density gradient can also be loaded dynamically before addition of the sample (Figs. 6-16H and 6-16I). To do this, the empty rotor is accelerated (Fig. 6-16H) and the gradient, beginning with the cushion or dense limit, is pumped into the loading ring of the distributor and is swept along the septa by centrifugal force. Thus it immediately forms a radial gradient. Following this, the sample to be fractionated is added as described earlier (i.e., beginning with the stage illustrated in Fig. 6-16C).

Sealed Operation (Fig. 6-17) Sealed operation of the TZ-28 rotor may optionally be used with superspeed centrifuges but is required with ultracentrifuges. For sealed operation, the standard distributor is replaced with the ultraspeed assembly. The rotor is mounted in the chamber of the ultracentrifuge, the stempiece inserted into the distributor, and the density gradient pumped into the stationary rotor light end first (Fig. 6-17A). With the gradient loaded, the stempiece is removed and equal amounts of sample loaded into the six sample channels (Fig. 6-17B). The sample flows into the rotor and forms a layer on the surface of the density gradient. Attachment of the sealing cover seals the rotor's contents.

The density gradient, together with the sample layer are reoriented from the vertical to the radial position (Figs. 6-17C and 6-17D) by slow acceleration of the rotor. Once reorientation is complete, the rotor is rapidly accelerated to the operating speed. From this point on, operation of the rotor is essentially the same as during unsealed use (i.e., Figs. 6-

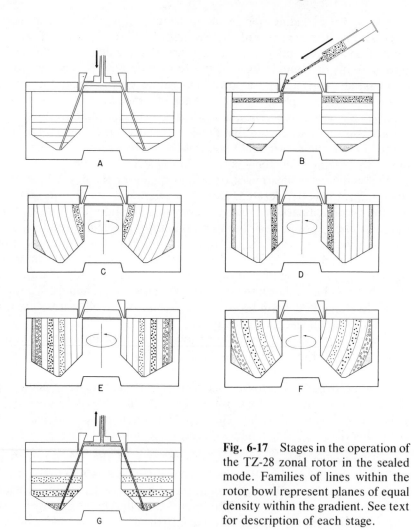

Fig. 6-17 Stages in the operation of the TZ-28 zonal rotor in the sealed mode. Families of lines within the rotor bowl represent planes of equal density within the gradient. See text for description of each stage.

17E, 6-17F, and 6-17G correspond to Figs. 6-16E, 6-16F, 6-16G, respectively).

In some applications it may be desirable to use an overlay so that buoyant particles (such as lipids) can move centripetally out of the sample zone during centrifugation. The overlay can be introduced into the rotor either statically or dynamically following introduction of the gradient and the sample. For flotation experiments, the sample is introduced last during static loading or first during dynamic loading. In this manner the sample zone will initially be located near the rotor wall; the buoyant particles will migrate toward the axis of rotation during centrifugation. As in the case of edge-unloading, rotating-seal zonal rotors, the TZ-28 may be temporarily decelerated during a run to remove the dense end of the gradient along with those particles that have already entered this region, the removed material replaced with new gradient, the rotor accelerated again, and the run continued.

OTHER REOGRAD ZONAL ROTORS

Most other reograd zonal rotors (as well as a modified version of the TZ-28, see Chapter 10) are used for continuous-flow collection of particles suspended in multiliter volumes and are rarely used for rate and/or isopycnic density gradient separations (Cline and Dagg, 1973; Nixon et al,, 1973). Reports have appeared in the literature from time to time describing modifications of B-type zonal rotors for reorienting gradient work to circumvent the deleterious effects that shearing has as fibrous particles or macromolecules (e.g., DNA) cross the faces of the rotating seal (Klucis and Lett, 1970). The only other commercial reograd zonal rotor of any consequence is the JCF-Z rotor manufactured by Beckman Instruments that is used in their superspeed centrifuges (Griffith and Wright, 1972). The JCF-Z has three interchangeable cores (see above), one of which is used for reorienting gradient work. In most respects the JCF-Z is similar to the SZ-14 and TZ-28 rotors, the density gradient and separated zones collected from funnel-shaped chambers formed by the core and rotor bowl floor. Unlike the SZ-14 and TZ-28 rotors, the JCF-Z is loaded statically only.

ROTOR-SPECIFIC APPLICATIONS OF A-TYPE, B-TYPE, AND REOGRAD ZONALS

A number of different factors should be weighed in the selection of a specific zonal rotor for a given application. Among the most important considerations are the sizes and the densities of the particles to be separated, but others include rotor and centrifuge availability and whether the particle separation is to be rate based or isopycnic. Although there is considerable overlap in the applications of various types of zonal rotors (see Fig. 6-18), some generalizations are possible. For example, B-type rotating-seal and reograd zonal rotors offer the highest centrifugal forces and can therefore be used for rate or isopycnic separations of small particles such as macromolecules and the smaller subcellular organelles. The lower operating speeds of A-type zonal rotors restrict their use to isopycnic and rate separations of somewhat larger particles.

Since both A-type and B-type rotating-seal zonal rotors are unloaded while they are spinning, the centrifugal force experienced by particles during the unloading interval may preclude the application of these rotors to rate separations of especially large particles such as whole cells. Rate separations of such large particles are usually carried out in shallow density gradients of low limiting densities and may require only a few minutes of centrifugation time. Introduction of the sample across the rotating-seal faces of a dynamically unloaded zonal rotor usually requires several minutes so that particles in that portion of the sample introduced first are well into the density gradient before the last portion of the sample has been added. Moreover, dynamic unloading takes about 30 minutes and is accompanied by continued particle sedimentation. In reograd zonals, the sample is either loaded at rest or quickly pumped into the spinning distributor; all particles in the sample begin sedimentation at essentially the same time. Since the rotor is decelerated to rest before unloading, particle sedimentation is effectively limited to the time at speed. Reograd

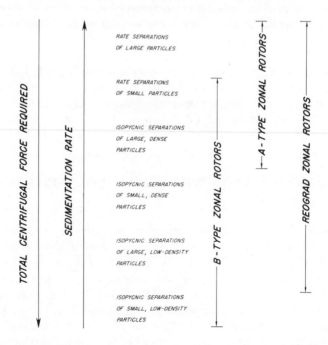

Fig. 6-18   Applications of various types of zonal rotor.

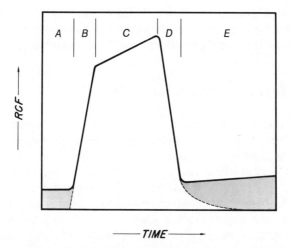

Fig. 6-19 Cumulative centrifugal force experienced by particles during centrifugation in dynamically loaded and unloaded zonal rotors and in reograd zonal rotors. In dynamically loaded and unloaded rotors, particles experience some additional centrifugal force during the loading and unloading intervals.

zonal rotors have been used effectively in very short runs to separate whole cells in very shallow density gradients (Sheeler and Doolittle, 1980). Figure 6-19 compares the cumulative centrifugal force experienced by particles in dynamically unloaded zonal rotors with that experienced in reograd zonal rotors.

SECTORIAL DILUTION

Although the sector-shaped compartments that characterize zonal rotors eliminate or at least reduce wall effects, they compound the difficulties of tailoring density gradients to achieve specific particle separations. In conventional centrifuge tubes the relationship between radius (i.e., distance from the axis of rotation) and volume is *linear,* but in the sector-shaped compartments of the zonal rotor the relationship is *exponential.* That is, the volume of a uniformly thick cylindrical zone increases in proportion to the square of the radius of the cylinder. Consequently, gradients in which the density changes linearly with respect to volume do *not* change linearly with respect to radius, and so forth. This effect is known generally as *sectorial expansion* or *sectorial dilution.* In Chapter 8 the relationships between volume, radius, and density in all major types of rotor used for density gradient separations are considered in depth, and therefore a discussion of the effects of sectorial dilution in sector-shaped zonal rotor compartments and the means devised for compensating or correcting for these effects is deferred until then.

THE CENTRIFUGAL FAST ANALYZER

Any discussion of the manifold contributions of Norman Anderson's MAN program to the field of centrifugal technology would not be complete without mention of Anderson's ingenious innovation called the "GeMSAEC" centrifugal fast analyzer (Hatcher and Anderson, 1969; Mashburn et al., 1970). ("GeMSAEC," pronounced gem-sack, is an acronym for "general medical sciences—atomic energy commission" and indicates the two major funding sources for the instrument's development—the National Institute of General Medical Sciences and the U.S. Atomic Energy Commission.) A complete discussion of the various types of GeMSAEC fast analyzer is beyond the scope and intent of this book, and the interested reader is referred to the review by Burtis et al. (1976) for a more thorough treatment of the subject than the one that follows.

Prior to the development of the centrifugal fast analyzer, virtually all automated systems designed for multisample clinical and biochemical assays were essentially mechanizations of manual techniques that analyze the samples in succession. That is, the reactions are *sequentially* initiated in each sample and the developing reactions of each sample are *sequentially* monitored. Anderson's GeMSAEC system represented a complete departure from this approach in that the individual reactions were *simultaneously* initiated in all samples and all samples were then *simultaneously* analyzed.

In Anderson's centrifugal fast analyzer, centrifugal force is used to simultaneously transfer and mix the biological samples and reagents together, after which the developing colorimetric reactions are simultaneously monitored in all mixtures. The heart of the system is the "multicuvette rotor" which contains a number of concentrically arranged sample and reagent wells or cavities, with the row of cavities along any radius of the rotor interconnected by narrow channels; some rotors contain as many as 42 rows. The outermost cavity of each row serves as a cuvette for spectrophotometric measurements.

The concentric wells of the rotor are initially filled with aliquots of the samples to be analyzed and the various reagent solutions. One or more rows is usually reserved for a reagent blank and controls. As the rotor is accelerated, the developing centrifugal force causes the aliquots to transfer radially from one cavity to another until they reach their respective cuvettes. The annular ring of cuvettes rotates through the light path of a spectrophotometric (or fluorometric) optical system, the light traversing each cuvette through its transparent upper and lower surfaces. Changes in optical density within each cuvette mixture are followed by synchronizing the spectrophotometer's output signals with the speed of the rotor. The changes in the optical properties of the contents of each cuvette are continuously compared with the reagent blank and are displayed on an oscilloscope as a series of spikes (one spike for each cuvette); the height of the spikes changes as the colorimetric reaction proceeds. The data may also be recorded as changing absorbance or transmittance values for each cuvette. Since the rotor spins at several thousand revolutions per minute, each cuvette is monitored many times each second; therefore, in effect, parallel changes are recorded in the optical properties of each sample-reagent mixture.

Certain commercial versions of the centrifugal fast analyzer provide for additional operational variations such as (1) loading of individual samples or reagents into the wells while the rotor is spinning, (2) application of bursts of air through successive cavities to assure thorough mixing, and (3) abrupt changes in rotor speed to resuspend packed particles during cell agglutination tests (Burtis et al., 1976).

It is to be noted that unlike the other centrifugal instruments and techniques described in this book, the centrifugal fast analyzer is not used to fractionate particle mixtures or to characterize particles on the basis of their sedimentation behavior.

REFERENCES AND RELATED READING

Books
Anderson, N. G., Ed. *The Joint National Institutes of Health—Atomic Energy Commission Zonal Centrifuge Program Semiannual Report,* July 1–December 31, 1962, ORNL-3415. Oak Ridge National Laboratory, Oak Ridge, Tennessee, 1962.

Anderson, N. G., Ed. *The Development of Zonal Centrifuges and Ancillary Systems for Tissue Fractionation and Analysis.* National Cancer Institute Monograph Series, No. 21, Bethesda, 1966.

Birnie, G. D., and Rickwood, D., Eds. *Centrifugal Separations in Molecular and Cell Biology.* Butterworths, London, 1978.

Dorvyl, G., Ed. *European Symposium of Zonal Centrifugation in Density Gradient.* Editions Cite Nouvelle, Paris, 1973.

Hinton, R. and Dobrota, M., *Density Gradient Centrifugation.* North-Holland, Amsterdam, 1976.

Reid, E., Ed. *Separations with Zonal Rotors*. Wolfson Bioanalytical Centre Publishers, London, 1971.

Reid, E., Ed. *Methodological Developments in Biochemistry*, Vol. 3, *Advances with Zonal Rotors*. Longmans Group Publishers, London, 1973.

Articles and Reviews

Albright, J., and Anderson, N. G. (1958) A method for rapid fractionation of particulate systems by gradient differential contrifugation. *Exp. Cell Res.*, **15**, 271.

Anderson, N. G. (1955) Studies on isolated cell components. VIII. High resolution gradient differential centrifugation. *Exp. Cell Res.*, **9**, 446.

Anderson, N. G. (1956) New Fractionation methods for isolating cellular proteins. *Bull. Amer. Phys. Soc.* **1** (Ser. II), 267.

Anderson, N. G., (1962) The zonal ultracentrifuge. A new instrument for fractionating mixtures of particles. *J. Phys. Chem.*, **66**, 1984.

Anderson, N. G. (1966) Zonal centrifuges and other separation systems. *Science* **154**, 103.

Anderson, N. G., Barringer, H. P., Babelay, E. F., and Fisher, W. D. (1964a) The B-IV zonal ultracentrifuge. *Life Sci.*, **3**, 667.

Anderson, N. G., Price, C. A., Fisher, W. D., Canning, R. E., and Burger, C. L. (1964b) Analytical techniques for cell fractions. IV. Reorienting gradient rotors for zonal centrifugation. *Anal. Biochem.*, **7**, 1.

Anderson, N. G., Barringer, H. P., Cho, N., Nunley, C. E., Babelay, E. F., Canning, R. E., and Rankin, C. T. (1966) The development of low-speed "A" series zonal rotors. In *The Development of Zonal Centrifuges and Ancillary Systems for Tissue Fractionation and Analysis. National Cancer Institute Monograph 21*, 113.

Anderson, N. G., Waters, D. A., Fisher, W. D., Cline, G. B., Nunley, C. E., Elrod, L. H., and Rankin, C. T. (1967) Analytical techniques for cell fractions. V. Characteristics of the B-XIV and B-XV zonal centrifuge rotors. *Anal. Biochem.*, **21**, 235.

Anderson, N. G., Rankin, C. T., Brown, D. H., Nunley, C. E., and Hsu, H. W. (1968) Analytical techniques for cell fractions. XI. Rotor B-XXIII—a zonal centrifuge rotor for center or edge unloading. *Anal. Biochem.* **26**, 415.

Anderson, N. G., Nunley, C. E., and Rankin, C. T. (1969a) Analytical techniques for cell fractions. XV. Rotot B-XXIX—a new high resolution zonal centrifuge for virus isolation and cell fractionation. *Anal. Biochem.*, **31**, 255.

Anderson, N. G., Waters, L. A., Nunley, C. E., Gibson, R. F., Schilling, R. M., Denny, E. C., Cline, G. B., Babelay, E. F., and Perardi, T. E. (1969b) K-series centrifuges. I. Development of the K-II continuous-sample-flow-with-banding centrifuge system for vaccine purification. *Anal. Biochem.* **32**, 460.

Burtis, C. A., Tiffany, T. O., and Scott, C. D. (1976) The Use of a centrifugal fast analyzer for biochemical and immunological analyses. In *Methods of* Biochemical Analysis, D. Glick, Ed. Wiley-Interscience, New York.

Cline, G. B., and Dagg, M. (1973) Particle separations in the J- and RK-types of flo-band zonal rotors. In *Methodological Developments in Biochemistry,* Vol. 3, *Advances with Zonal Rotors,* E. Reid, Ed. Longmans Group Publishers, London.

Elrod, L. H., Patrick, L. C., and Anderson, N. G. (1969) Analytical techniques for cell fractions, XII. Rotor A-XVI, a plastic gradient reorienting rotor for isolating nuclei. *Anal. Biochem., 30,* 230.

Griffith, O. M., and Wright, H. (1972) Resolution of components from rat liver homogenate in reorienting density gradients. *Anal. Biochem., 47,* 575.

Hatcher, D. W., and Anderson, N. G. (1969) GeMSAEC: A new analytical tool for clinical chemistry. Total serum protein with the Biuret reaction. *Amer. J. Clin. Pathol., 52,* 645.

Klucis, E. S., and Lett, J. T. (1970) Zonal centrifugation of mammalian DNA. *Anal. Biochem., 35,* 480.

Mashburn, D. N., Stevens, R. H., Willis, D. D., Elrod, L. H., and Anderson, N. G. (1970) Analytical techniques for cell fractions. XVII. The G-IIC fast analyzer system. *Anal. Biochem., 35,* 98.

Nixon, J. C., McCarty, K. S., and McCarty, K. S. (1973) The use of a reorienting density gradient rotor with continuous sample flow for the isolation of calf thymus nuclei. *Anal. Biochem., 55,* 132.

Reimer, C. B., Baker, R. S., van Frank, R. M., Newlin, T. E., Cline, G. B., and Anderson, N. G. (1967) Purification of large quantities of influenza virus by density gradient centrifugation. *J. Virol., 1,* 1207.

Sheeler, P. (1971) Reorienting density gradient zonal centrifugation. *Amer. Lab.* (February, p. 19).

Sheeler, P. (1974) Reorienting density gradients and the SZ-14 rotor. In *Methodological Developments in Biochemistry,* Vol. 4, E. Reid, Ed. Longmans Group Publishers, London, p. 47.

Sheeler, P., and Wells, J. R. (1969) A reorienting gradient zonal rotor for low-speed separation of cell components. *Anal. Biochem., 32,* 38.

Sheeler, P., Gross, D. M., and Wells, J. R. (1971) Zonal centrifugation in reorienting density gradients. *Biochim. Biophys. Acta, 237,* 28.

Sheeler, P., Gross, D. M., and Wells, J. R. (1971) Zonal centrifugation in reorienting density gradients. *Biochim. Biophys. Acta, 237,* 28.

Sheeler, P., and Wells, J. R. (1973) Zonal Centrifugation using reorienting density gradients. *Eur. Symp. Zonal Centrifugation,* 259.

Sheeler, P. and Doolittle, M. H. (1980) Separation of mammalian cells by velocity sedimentation. *Amer. Lab., 12* (4), 49.

Supplementary References to Illustrative Applications of A-Type, B-Type, and Reorienting Gradient Zonal Rotors

A-Type Zonal Rotors

Barber, M. L., and Canning, R. E. (1966) Extraction of contractile proteins from myofibrils prepared by rate-zonal centrifugation. *Nat. Cancer Inst. Monogr.*, **21**, 345.

Evans, W. H. (1971), Large scale preparation and characterisation of liver plasma membranes, in *Separations with Zonal Rotors* (E. Reid, Ed.), Wolfson Bioanalytical Laboratories Publishers, London.

Evans, W. H., and Gurd, J. W. (1973) Properties of a 5'-nucleotidase purified from mouse liver plasma membranes. *Biochem. J.*, **133**, 189.

Hilderson, H. J. (1974) Study on bovine thyroid nuclei isolated with an A-XII zonal rotor. In *Methodological Developments in Biochemistry*, Vol. 4, *Subcellular Studies*, E. Reid, Ed. Longmans Group Publishers, London.

Johnston, I. R., Mathias, A. P., Pennington, F., and Ridge, D. (1968) The fractionation of nuclei from mammalian cells by zonal centrifugation. *Biochem. J.* **109**, 127.

Kung, F. C., and Glaser, D. A. (1977) Synchronization of *Escherichia coli* by zonal centrifugation. *Appl. Environ. Microsc.*, **34**, 328.

Neal, W. K., Hoffmann, H. P., and Price, C. A. (1971) Sedimentation behavior and ultrastructure of mitochondria from repressed and derepressed yeast, Saccharomyces cerevisiae. *Plant Cell Physiol.*, **12**, 181.

Poole, R. K. and Lloyd, D. (1974) Use of zonal rotors in the analysis of the cell-cycle in yeasts. In *Methodological Developments in Biochemistry*, Vol. 4, *Subcellular Studies*, (E. Reid, Ed.) Longmans Group Publishers, London.

Price, C. A. (1973) Separation of chloroplasts by isopycnic, rate-zonal, and continuous-flow density gradient centrifugation. In *European Symposium of Zonal Centrifugation in Density Gradient*, G. Dorvyl, Ed. Editions Cite Nouvelle, Paris.

Price, C. A., and Hirvonen, A. P. (1967) Sedimentation rates of plastids in an analytical zonal rotor. *Biochim. Biophys. Acta*, **148**, 531.

Price, M. R., Harris, J. R., and Baldwin, R. W. (1973) Isolation of nuclear "ghosts" from normal rat liver hepatoma. In *Methodological Developments in Biochemistry*, Vol. 3, *Advances with Zonal Rotors*, (E. Reid, Ed.) Longmans Group Publishers, London.

Schneider, E. L., and Salzman, N. P. (1970) Isolation and zonal fractionation of metaphase chromosomes from human diploid cells. *Science*, **167**, 1141.

Schuel, H., Schuel, R., and Unakar, N. J. (1968) Separation of rat liver lysosomes and mitochondria in the A-XII zonal centrifuge. *Anal. Biochem.*, **25**, 146.

Still, C. C., and Price, C. A. (1967) Bulk Separation of chloroplasts with intact membranes in the zonal centrifuge. *Biochim. Biophys. Acta*, **141**, 176.

Warmsley, A. M. H., and Pasternak, C. A. (1970) The use of conventional and zonal centrifugation to study the life cycle of mammalian cells. *Biochem. J.*, **119**, 493.

B-Type Zonal Rotors

Bloemendal, H., Berns, T., Zweers, A., Hoenders, H., and Benedetti, E. L. (1972) The state of aggregation of alpha-crystallin detected after large-scale preparation by zonal centrifugation. *Eur. J. Biochem.*, **24**, 401.

Bretz, U., and Baggliolini, M. (1973) Resolution of three distinct populations of nerve endings from rat brain by zonal isopycnic centrifugation. In *European Symposium of Zonal Centrifugation in Density Gradient*, G. Dorvyl, Ed. Editions Cite Nouvelle, Paris.

Brown, D. H. (1968) Separation of mitochondria, peroxisomes and lysosomes by zonal centrifugation in a Ficoll gradient. *Biochim. Biophys. Acta*, **162**, 152.

Brown, D. H., Carlton, E., Byrd, B., Harrell, B., Harrell, B., and Hayes, R. L. (1973) A rate-zonal centrifugation procedure for screening particle populations by sequential product recovery utilizing edge-unloading zonal rotors. *Arch. Biochem. Biophys.*, **155**, 9.

Cameron, I. L., Griffin, E. E., and Rudick, M. J. (1971) Macromolecular events following refeeding of starved *Tetrahymena*. *Exp. Cell Res.*, **65**, 265.

Cox, R. A., and Pratt, H. (1973) Fractionation of subribosomal particles by zonal ultracentrifugation. In *Methodological Developments in Biochemistry*, Vol. 3, *Advances with Zonal Rotors*, E. Reid, Ed. Longmans Group Publishers, London.

Eikenberry, E. F., Bickle, T. A., Traut, B. R., and Price, C. A. (1970) Separation of large quantities of ribosomal sununits by zonal ultracentrifugation. *Eur. J. Biochem.* **12**, 113.

Fleischer, B., Fleischer, S., and Ozawa, H. (1969) Isolation and characterization of Golgi membranes from bovine liver. *J. Cell Biol.*, **43**, 59.

Hall, W. T., and Bond, H. E. (1973) Purification and characterization of the hepatitis-associated (HB-Ag) antigen. In *European Symposium of zonal Centrifugation in Density Gradient*, G. Dorvyl, Ed. Editions Cite Nouvelle, Paris.

Hinton, R. H., Dobrota, M., Fitzsimons, J. T. R., and Reid, E. (1970) Preparation of plasma membrane fraction from rat liver by zonal centrifugation. *Eur. J. Biochem.*, **12**, 349.

Lee, T., Swatrzendruber, D. C., and Snyder, F. (1969) Zonal centrifugation of microsomes from rat liver: Resolution of rough- and smooth-surfaced membranes. *Biochem. Biophys. Res. Commun.*, **36**, 748.

Mahaley, M. S., Day, E. D., Anderson, N., Wilfong, R. F., and Brater, C. (1968) Zonal centrifugation of adult, fetal, and malignant brain tissue. *Cancer Res.*, **28**, 1783.

Taylor, D. G., and Crawford, N. (1974) The subcellular fractionation of pig blood platelets by zonal centrifugation. In *Methodological Developments in Biochem-*

istry, Vol. 4, *Subcellular Studies*, E. Reid, Ed. Longmans Group Publishers, London.

Trelease, R. N., Becker, W. M., Gruber, P. J., and Newcomb, E. H. (1971) Microbodies (glyoxysomes and peroxisomes) in cucumber cotyledons. *Plant Physiol.*, **48**, 461.

Weaver, R. A., and Boyle, W. (1969) Purification of plasma membranes of rat liver. *Biochim. Biophys. Acta*, **173**, 377.

Wilcox, H. G., Davis, D. C., and Heimberg, M. (1969) The isolation of lipoproteins from human plasma by ultracentrifugation in the Ti-14 and Ti-15 zonal rotors. *Biochim. Biophys. Acta*, **187**, 147.

Williamson, R. (1969) Purification of DNA by isopycnic banding in CsCl in a zonal rotor. *Anal. Biochem.*, **32**, 158.

Reorienting Gradient Zonal Rotors

Carter, C. E., Wells, J. R., and Macinnis, A. J. (1972) DNA from anaerobic adult *Ascaris lumbricoides* and *Hymenolepis diminuta* isolated by zonal cnetrifugation. *Biochim. Biophys. Acta*, **262**, 135.

Childress, W. J., Freedman, R. I., Koprowski, C., Doolittle, M. H., Sheeler, P., and Oppenheimer, S. B. (1979) Surface characteristics of separated subpopulations of mouse teratocarcinoma cells. *Exp. Cell Res.*, **122**, 39.

Cummins, J. E., and Day, A. W. (1973) Cell cycle regulation of mating type alleles in the smut fungus *Ustilago violacea*. *Nature*, **245**, 259.

Day, A. W., and Cummins, J. E. (1973) Temporal allelic interaction, a new kind of dominance. *Nature*, **245**, 260.

Gross, D. M., and Barajas, L. (1975) The large-scale isolation of rennin-containing granules from rabbit renal cortex by zonal centrifugation. *J. Lab. Clin. Med.*, **85**, 467.

Neurath, A. R., Cosio, L., Prince, A. M., and Lippin, A. (1973) Purification of hepatitis B antigen associated particles: Use of a reorienting gradient rotor. *Proc. Soc. Exp. Biol. Med.*, **144**, 384.

Pickett, C. B., Cascarano, J., and Johnson, R. (1977) Oxidative phosphorylation in rat liver mitochondria isolated by rate zonal centrifugation: Examination of Ficoll gradients and subpopulations of mitochondria. *J. Bioenerg. Biomem.*, **9**, 271.

Smuckler, E. A., Riddle, M., Koplitz, M., and Glomset, J. (1974) The use of a reorienting zonal rotor for cell membrane isolation. In *Methodological Developments in Biochemistry*, Vol. 4, *Subcellular Studies*, E. Reid, Ed. Longmans Group Publishers, London.

Stocco, D. M., Cascarano, J., and Wilson, M. A. (1977) Quantitation of mitochondrial DNA, RNA and protein in starved and starved-refed rat liver. *J. Cell. Physiol.*, **90**, 295.

Wells, J. R. (1974) Mitochondrial DNA synthesis during the cell cycle of *Saccharomyces cerevisiae*. *Exp. Cell Res.*, **85**, 278.

Wells, J. R., Sheeler, P., and Gross, D. M. (1972) A reorienting density gradient rotor for zonal centrifugation. *Anal. Biochem.*, **46**, 7.

Wells, J. R., and James, T. W. (1972) Cell cycle analysis by culture fractionation. *Exp. Cell Res.*, **75**, 465.

Wells, J. R., and Sheeler, P. (1973) The separation of cells and cell components in the SZ-14 rotor. In *European Symposium of Zonal Centrifugation in Density Gradient*, G. Dorvyl, Ed. Editions Cite Nouvelle, Paris.

Wells, J. R., Opelz, G., and Cline, M. J. (1977) Characterization of functionally distinct lymphoid and myeloid cells from human blood and bone marrow. I. Separation by a buoyant density gradient technique. *J. Immunol. Methods,* **18**, 63.

Wells, J. R., Opelz, G., and Cline, M. J. (1977) Characterization of functionally distinct lymphoid and myeloid cells from human blood and bone marrow. II. Separation by velocity sedimentation. *J. Immunol. Methods,* **18**, 79.

Wilson, M. A., and Cascarano, J. (1972) Biochemical heterogeneity of rat liver mitochondria separated by rate zonal centrifugation. *Biochem. J.,* **129**, 209.

Wilson, M. A., Cascarano, J., Wooten, W. L., and Pickett, C. B. (1978) Quantitative isolation of liver mitochondria by zonal centrifugation. *Anal. Biochem.* **85**, 255.

SEVEN

Recent Innovations in Density Gradient Methodology

Notwithstanding the tremendous progress made between 1950 and 1970 in centrifuge and rotor design and capability and in methodological approaches to cellular and subcellular fractionation, further changes, improvements, and novel additions to the technology of this area continue to be made by the manufacturers of centrifuges and accessory instruments and also by independent investigators striving to attack old problems in new ways. Some recent innovations in density gradient methodology are described in this chapter.

VERTICAL TUBE ROTORS

What is probably the most recent and also the most surprising innovation in density gradient methodology is the *vertical tube* rotor (Fig. 7-1). In many ways, the operation of vertical tube rotors and the prerequisites for their effective use are like those of reorienting gradient zonal rotors (see Chapter 6). Indeed, the original development of the vertical tube rotor by Du Pont Instruments is clearly based on their experiences with their reograd zonal rotors.

Basically, a vertical tube rotor is a fixed-angle rotor in which the angle has been reduced to zero degrees; that is, the recesses in the rotor that accept the centrifuge tubes are *parallel* to the rotor's axis of rotation. This, however, raises special sealing problems for the centrifuge tubes, for unlike fixed-angle rotors in which the tube cap experiences little or no hydrostatic pressure (i.e., the meniscus formed by the tube's contents lies centrifugal to the cap), the tube caps in a vertical tube rotor are

Fig. 7-1 Vertical tube rotors for ultraspeed and superspeed centrifuges. (Courtesy of E. I. DuPont and Company.)

subjected to appreciable upward pressure. As a result, a special sealing mechanism is required to prevent tube leakage at speed. Also, unlike fixed-angle rotors, vertical tube rotors are used almost exclusively for density gradient separations and not for pelleting particles.

Operation of a Vertical Tube Rotor   The operation of a vertical tube rotor is depicted diagrammatically in Fig. 7-2. The centrifuge tube is loaded with the density gradient and the sample to be fractionated carefully layered on top. The tube is sealed, placed in the rotor, and the rotor slowly accelerated. During acceleration, the gradient and the sample undergo "reorientation"—a phenomenon described previously in connection with density gradient centrifugation in fixed-angle rotors (Chapter 5) and reograd zonal rotors (Chapter 6). When the tube is initially inserted into the rotor, each isodense plane in the gradient takes the form of a circle, but as the rotor is accelerated, each isodense circle is transformed into a small segment of a paraboloid of revolution, the focus of which lies on the axis of rotation. The paraboloids become steeper as the rotor gains speed and eventually (usually by 1000 rpm) reach verticality. At this point the density gradient is distributed across the diameter of the centrifuge tube,

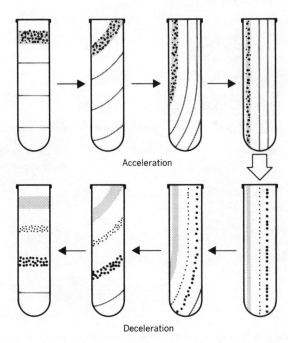

Acceleration

Deceleration

Fig. 7-2 Stages in the operation of a vertical tube rotor for density-gradient separations of particles. The axis of rotation is to the left of the tubes. See text for details.

the sample taking the form of a tall narrow zone at the centripetal side. Once reorientation is complete, the rotor may be more quickly accelerated to the operating speed. Particles in the sample sediment through the gradient to form a number of separate zones. Once the separation of particles is achieved, the rotor may be decelerated and the gradient reoriented to its original distribution in the tube. Slow and smooth deceleration of the rotor below 1000 rpm is important in order to prevent horizontal swirling of the gradient and separated particle zones. The tube is then removed from the rotor and the gradient and separated particles collected as a series of fractions.

    Superspeed and ultraspeed vertical tube rotors are now produced commercially by several companies, including DuPont/Sorvall Instruments, Beckman Instruments and Measuring and Scientific Equipment, Ltd. (see Appendixes). The slow acceleration–deceleration requirement for smooth and gradual reorientation of the gradient in the rotor's tubes is a feature that is either built into or may be added to the companies' more recent lines of centrifuges.

Sealing the Centrifuge Tubes of a Vertical Tube Rotor   Two alternative procedures are available for solving the problem of affecting a seal at the top of the centrifuge tube that is able to resist leakage at high rotor speed. In one approach, which employs a special cap assembly that becomes part of the rotor body, the tubes are sealed *after* they are placed in the rotor. In the other approach, specially designed tubes (see below) are "heat-sealed" *before* they are placed in the rotor. A spacer cap attached to the rotor body above each tube serves to locate and secure each tube in position but does not affect a seal.

Figure 7-3 shows the basic tools and components used to seal the open ends of conventional centrifuge tubes after they are positioned in the cavities of a vertical tube rotor. A tube plug is carefully inserted into the open end of each tube without disturbing the density gradient or sample

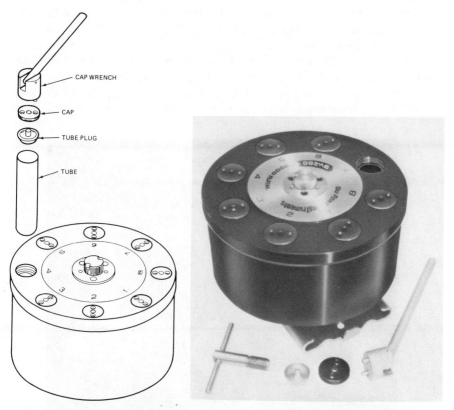

Fig. 7-3   Basic components of a vertical tube rotor. The photograph to the right shows the T-bar carrying handle and the vise used to secure the rotor when the tube caps are tightened. (Courtesy of E. I. DuPont and Company.)

zone. A threaded cap is then lowered over each plug and slowly screwed into the rotor body by using a cap wrench. For ultraspeed vertical tube rotors, a torque wrench set to exert precisely the amount of downward pressure necessary to seal the upper end of the centrifuge tube is used. As the tube plug is driven into the upper end of the centrifuge tube, its flared edge acts to spread the end of the centrifuge tube outward and against a correspondingly tapered surface at the top of each cavity in the rotor body. Compressed between these two flared surfaces, the·end of the centrifuge tube is tightly sealed. A rotor holding fixture mounted on a laboratory bench or other horizontal surface prevents the rotor from turning as the tube caps are tightened (or loosened). The assembled and completely sealed rotor is then transferred to the centrifuge.

After centrifugation and removal of the tubes from the rotor, the deformed rims of the tubes may be trimmed away by using a special trimmer accessory. The tube may then be unloaded more readily using a density gradient fractionator (Chapter 4). Each centrifuge tube can be used only once in a vertical tube rotor.

Motivated by the need for an alternative and perhaps simpler method of sealing the tubes for vertical tube rotors, Beckman Instruments in 1978 introduced "quick-seal" tubes (Fig. 7-4). As it turns out, the tubes may be used with fixed-angle rotors as well as with vertical tube rotors. The

Fig. 7-4 Quick-seal tubes and spacers. (Courtesy of Beckman Instruments, Inc.)

upper end of each quick-seal tube takes the form of a hemisphere with a short, central inlet section. After the tube is filled with the density gradient and the sample (using a narrow cannula or a length of flexible tubing inserted through the inlet), a small metal seal former is slipped over the inlet and the tube is then placed in a sealing device. Heat is applied to the seal former for about 15 sec, and this acts to fuse the inlet closed. After this, the seal former is removed, the tube placed in the centrifuge rotor, and a reusable spacer placed between the tube and the rotor lid or recess cap (see Fig. 7-4). When centrifugation is completed, the tube is punctured to allow recovery of the gradient and separated particles.

Also available for use with quick-seal tubes is a sample application device (Fig. 7-5). After the tube is loaded with the density gradient, it is placed in the device and gentle pressure used to compress the tube wall, thereby forcing the light end of the gradient through the tube inlet and into the neck of a small plastic funnel. The sample to be fractionated is then layered directly on the gradient in the funnel. When the pressure is gradually released, the gradient and the sample are drawn back into the tube, and the tube is then ready to be sealed.

Origins of Vertical Tube Rotors   Vertical tube rotors were introduced in 1974 and their special effectiveness (see below) for rate and isopycnic

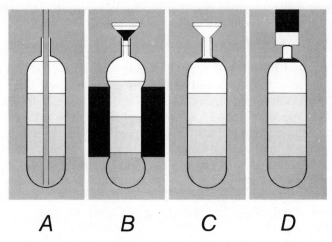

Fig. 7-5 Sample application device for quick-seal tubes: *A*, density gradient loaded through cannula; *B*, tube gently compressed to displace light end of gradient into sample loading funnel; *C*, pressure released to draw light end of gradient and sample into tube; *D*, sealing cap placed over tube insert and heat applied to obtain seal. (Courtesy of Beckman Instruments, Inc.)

separations immediately appreciated. In view of the simplicity of the basic idea behind this type of rotor, it is surprising that this approach to density gradient centrifugation was not attempted in the early 1960s, when isopycnic banding was first shown to be feasible in fixed-angle rotors; for in fixed-angle rotors, the same type of gradient and zone redistribution within the centrifuge tube occurs during rotor acceleration and deceleration. A little-known study carried out by A. V. Masket, who was working in the laboratory of J. W. Beams, and published more than 30 years before the initial vertical tube rotors were produced commercially describes what is in effect the first such design (see Masket, 1941).

In the late 1930s and the early 1940s, a number of different laboratories examined and compared the effectiveness of various configurations of experimental ultraspeed fixed-angle rotors for concentrating or pelleting particle suspensions. It was recognized that although centrifugal "efficiency" (see Chapter 9) increases as the value of the fixed angle is diminished, if the meniscus is to remain radial to the lip of the centrifuge tube at speed, there is a corresponding reduction in the usable space within the tube. In fixed-angle rotors that have a small angle of inclination, this space was occupied by a solid tube insert to prevent tube collapse. Angles less than 20° were generally considered to forfeit too large a portion of the rotor capacity to be practical. Using an altogether different and new approach to rotor design, Masket constructed a series of rotors in which the angle of inclination was reduced to 10°. To retain the full capacity of the centrifuge tubes while simultaneously taking advantage of the increased pelleting efficiency of so small an angle of inclination, Masket replaced the rotor lid with a series of caps that screwed down into each tube recess and that forced a tapered plug into each plastic centrifuge tube. A seal was thus achieved in what is essentially the same manner as that shown in the vertical tube rotor in Fig. 7-3. Masket's design worked faultlessly at speeds of up to 57,000 rpm, and although he did not construct a rotor having an angle of inclination of zero degrees, he suggested that his design made such a rotor feasible. At this early stage in the evolution of centrifugal methodology there would have been no motivation for a zero-degree fixed-angle rotor (i.e., a vertical tube rotor), for density gradient centrifugation as a means of particle fractionation was not introduced until the 1950s and Masket's goals were to improve pelleting efficiency.

Effectiveness of Vertical Tube Rotors   For a combination of reasons, vertical tube rotors are particularly effective for density gradient centrifugation, in many instances yielding separations in considerably less time than achieved in other rotors operating either at the same speed or higher speeds (Gregor, 1977; Wells and Brunk, 1979). This may be explained by examining Fig. 7-6, which shows the radial distance to the meniscus of loaded centrifuge tubes in swinging-bucket, fixed-angle, and vertical tube rotors of the same maximum radius. Clearly, at equivalent rotational speeds, even the minimum RCF experienced by particles in a vertical tube rotor is appreciably greater than that in the other two rotor types. As a result, particles will reach their final positions in the density gradient in less time in a vertical tube rotor. Referring again to Fig. 7-6, it can also be seen that even if the same density gradient initially fills each of the three centrifuge tubes, the radial distance necessarily traversed by any population of sedimenting particles to reach a given density plane in the gradient is going to be smallest in the vertical tube rotor. Yet, although the radial distance separating successive particle zones in a vertical tube rotor at speed may be quite small, this distance is increased as the gradient and the zones are reoriented from the radial to the vertical position.

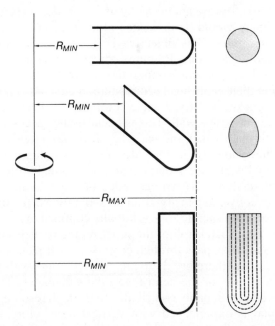

Fig. 7-6 Minimum and maximum radii in swinging-bucket (top), fixed-angle (middle), and vertical tube (bottom) rotors. Note the short path length and the high R_{min} value of the vertical tube rotor. Also shown are the cross-sectional areas of isodense planes traversed by the sedimenting particles.

Unlike swinging-bucket rotors, vertical tube rotors and fixed-angle rotors have no moving parts and are machined from a single block of aluminum or titanium. As a consequence, these rotors are generally more stable, can be operated safely at higher speeds, and have more tube positions than do swinging-bucket rotors of comparable tube volume. Some vertical tube rotors can be operated at speeds of 75,000 to 80,000 rpm (see Appendix), producing RCF values in excess of 400,000g, *even at the minimum tube radius!*

The efficiency of a rotor either for pelleting particles or sedimenting them a specific radial distance through a density gradient is a measure that can be ascertained if the size, the shape, and the density of the particles and the geometric specifications and operating speed of the rotor are known. These efficiencies are known as rotor K and K' values and can be used to advantage in planning experiments and selecting the most appropriate rotor. The efficiencies of vertical tube rotors, fixed-angle rotors, and swinging-bucket rotors are considered and compared in such terms in Chapter 9.

Vertical tube rotors are not normally used to pellet suspended particles since the sediment would be deposited along the entire centrifugal wall of the centrifuge tube. This objective is deferred instead to swinging-bucket and fixed-angle rotors. Rather, vertical tube rotors are used almost exclusively for density gradient separations. The radial paths followed by particles during their centrifugal sedimentation demands that in the tubes of a swinging-bucket and fixed-angle rotor the particles will eventually encounter the lateral tube wall. This is because the successive cross-sectional planes through which the particles travel are circles or ellipses of unchanging dimensions (Fig. 7-6). Moreover, as noted earlier (Chapter 5), the additional wall effects that occur in fixed-angle rotors as a result of the downward-sloping centrifugal edge of the tube limits the use of these rotors to *isopycnic* density gradient separations. In contrast, in the vertical tube rotor the sample zone is initially confined to a narrow *vertical* plane near the centripetal wall of the tube. As the particles in the sample sediment radially, they pass through cross-sectional planes of increasing width (Fig. 7-6), and therefore there are no encounters with the tube wall in the centripetal half of the centrifuge tube's diameter. This reduction in wall effects unquestionably contributes to the high resolution of particle populations that is experienced with vertical tube rotors. Indeed, it is possible to perform *both* rate and isopycnic density gradient separations of whole cells (Sheeler and Doolittle, 1979), subcellular organelles and other particles (Gregor, 1977; Hofman et al., 1978), and macromolecules (Jordon and Prestwich, 1977; Wells and Brunk, 1979) in vertical tube rotors.

PRODUCING MULTIPLE DENSITY GRADIENTS

Over the years, a number of methods have been proposed and a number of interesting devices constructed for *simultaneously* producing a multiplicity of identical density gradients in centrifuge tubes for parallel runs. In this way, the results of density gradient separations of several different samples can be reliably compared since both the conditions of centrifugation and the gradient employed are the same for each sample. Parallel runs in identical gradients are also desirable when it is necessary to pool the corresponding fractions obtained from different samples. The alternative is the time-consuming and laborious chore of producing the gradients successively. With the increase from 3 to 6 in the number of bucket positions available in ultraspeed swinging-bucket rotors and with the advent of 8- to 16-position vertical tube rotors, effective methods for quickly producing multiple identical density gradients have become all the more important.

Multiple Self-generating Gradients   A number of density gradient solutes automatically form density gradients if a uniform solution of the solute is subjected to centrifugation for some time. These include CsCl (Flamm et al., 1969; Brunk and Leick, 1969), metrizamide (Hell et al., 1974), Ludox (Pertoft and Laurent, 1969) and Percoll (Pertoft et al., 1978). (The physical bases underlying the automatic formation of density gradients by centrifugation were considered in Chapter 5.) In contrast to the behavior of CsCl, solutions of metrizamide, Ludox, and Percoll generate usable density gradients at moderate rotor speeds and in reasonable lengths of time if the centrifugation is carried out in a fixed-angle rotor, but not if carried out in swinging-bucket or zonal rotors. Among other influential factors, the shape of the gradient that is automatically generated depends on the speed and time of centrifugation, the angle of inclination of the rotor, and the operating temperature. For example, at 5°C a 1.15-gm/ml solution of metrizamide centrifuged at 35,000 rpm for 40 hr in an MSE 10 × 10 fixed-angle rotor (35° angle of inclination) produces a gently concave gradient varying in density from 1.06 to 1.35 gm/ml. Higher-molecular-weight and polydisperse solutes like Ludox and Percoll form usable gradients in much shorter times (as little as 30 min) and at much lower rotor speeds (15,000–20,000 rpm). In general, the higher the speed and the longer the rotor spins, the steeper the resulting density gradient. Guides to gradient shapes produced using various solutes in various rotors are to be found in the literature cited at the end of the chapter. Thus, if one chooses centrifuge tubes that can be used interchangeably in specific models of swinging-bucket, fixed-angle, and vertical tube rotors, as many as 12 identical density gradients can simultaneously be generated using the fixed-angle rotor, the samples then layered onto each gradient, and the tubes then used in either a swinging-bucket or a vertical tube rotor to achieve the parallel separation.

Multiple Preformed Gradients   Although a number of gradient makers available commercially strive to do so (see Fig. 4-4 for an example), a multiplicity of identical density gradients cannot be reliably formed by simply subdividing a single gradient stream into a number of channels leading to separate centrifuge tubes. Inevitable variations in the flow rates into each tube result in unequal gradient volumes. Depending on the extent of the variation, the resulting imbalance in an ultraspeed rotor could significantly harm the centrifuge drive. Static or rotating manifolds intended to more effectively apportion a single gradient source to an array of centrifuge tubes or channels are fraught with problems due to the unavoidable presence of small variations (in the dimensions or the properties) in the channels of the manifold or in the connecting lines. Exotic devices using multiple peristaltic pump heads produce variable flow rates depending on tubing conditions, wall uniformity, roller wear, and so on.

N. G. Anderson (see Candler et al., 1967) and Siaskotos and Wirth (1967) independently described modified centrifuge rotors in which an axially mounted distributor spinning at high speed with the rotor received a single gradient stream and then apportioned it equally to the rotor's tubes. Anderson's design employed a 12-position fixed-angle rotor, whereas a 12-position swinging-bucket rotor was used by Siaskotos and Wirth. Both instruments produced parallel gradients that were virtually identical; however, the production of a commercial version using their principles has never been undertaken.

Beckman Instruments manufactures a density gradient maker that is capable of producing three identical gradients at a time having volumes of up to 60 ml per gradient (Fig. 7-7). In this instrument, three pairs of syringes are loaded with the gradient's light and dense limits, and with a pair of cam-driven plungers, the limits are forced *in continuously varying proportions* through three small mixing chambers and into the centrifuge tubes. The cams are shaped so that the resulting gradient is linear in density with respect to volume. The need for a larger number of gradients could be met by using more than one gradient former, although modification of the basic design so that the instrument accepted additional pairs of syringes and mixing chambers should be feasible. The key to the success of the Beckman instrument stems from the fact that its design circumvents the difficulty of driving *one* liquid stream of continuously chang-

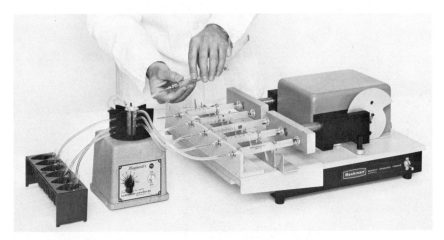

Fig. 7-7 Multiple density-gradient maker (see text for description). (Courtesy of Beckman Instruments, Inc.)

ing density into each of the lines leading to the centrifuge tubes *at precisely the same rate*.

An apparatus used in the author's laboratory and that produces up to eight identical gradients is shown in Fig. 7-8; it consists of (1) specially designed centrifuge tubes, and (2) a device for loading and unloading the tubes (Sheeler et al., 1978). In contrast to the principle used in the Beckman gradient former, uniform and identical gradients in each tube are assured by simultaneously *drawing* the gradients into the tubes rather than *driving* the gradients into the tubes. This is achieved by simultaneously lowering the plungerlike floors of each tube (Fig. 7-9). Since all tube floors can be lowered at precisely the same rate and since uniform lowering is necessarily accompanied by uniform filling, each tube fills in exactly the same way. Collective unloading of the tubes at the conclusion of a run (if, e.g., it is necessary to pool fractions) is achieved by elevating all tube floors simultaneously. Separate unloading of each gradient requires only that negative pressure be applied to the top of each tube, thus drawing the tube floor upward, and displacing the density gradient and the entrained zones of particles (Fig. 7-10).

Two types of tube are used with the apparatus: modified plastic syringe barrels and modified polyallomer centrifuge tubes. To use syringes, the barrel is shortened and the rubber plunger mounted on a disk; when inserted into the bottom of the barrel, the plunger and the disk form a seal. Since the bottoms of rotor buckets are hemispherical, the syringe barrels

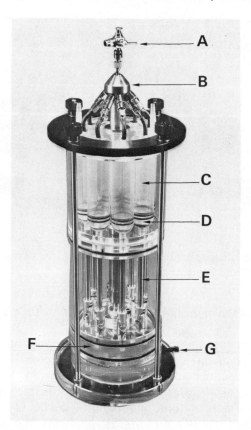

Fig. 7-8 Apparatus for producing up to eight identical gradients in modified centrifuge tubes: A, gradient (and sample) inlet control valve; B, distributor; C, modified centrifuge tubes (see also Fig. 7-9); D, tube plunger inserts; E, plunger rods; F, piston for raising and lowering plungers; G, water inlet for raising and lowering piston.

are mounted on hemispherical supports prior to centrifugation. The modified polyallomer centrifuge tubes are fitted with molded inserts that serve as plungers and also act to seal an opening in the base of the tubes. A conical cap is seated in the open end of the centrifuge tube during gradient loading and unloading. The syringe or centrifuge tube plungers are connected through rods to a piston. If the piston is elevated (or lowered), the plungers are raised (or lowered) in concert.

To simultaneously load the tubes, the plungers are slowly pulled downward by the piston, thereby drawing the gradient (dense end first) into each of the tubes through a distributor mounted above. The gradient en-

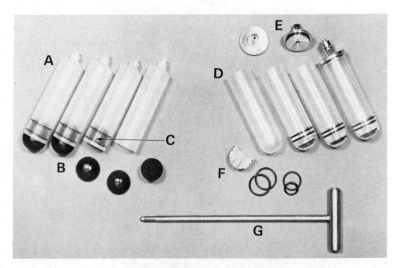

Fig. 7-9 Modified syringe barrels (left) and centrifuge tubes (right) used with the apparatus shown in Fig. 7-8: *A*, syringe barrels; *B*, hemispherical barrel supports; *C*, plungers; *D*, modified centrifuge tubes; *E*, conical loading and unloading caps; *F*, tube inserts; *G*, insert extractor.

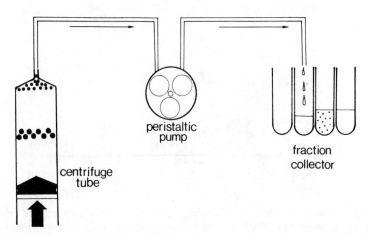

Fig. 7-10 Recovery of density gradients from special centrifuge tubes. Gradient and particle zones are displaced upward and out of the tube under negative pressure by using a peristaltic pump. No displacing solution is necessary.

153

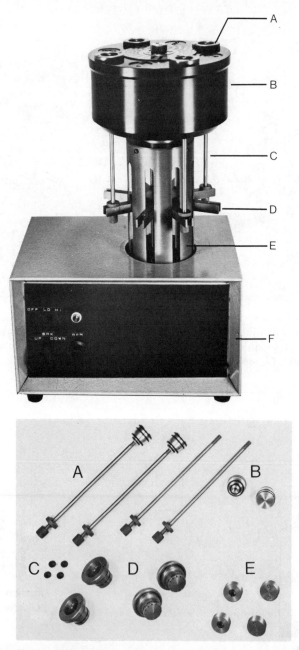

Fig. 7-11 Parallel loading (and unloading) vertical tube rotor (see text for details). *Top*: *A*, gradient–sample feed device; *B*, rotor body; *C*, rods for raising and lowering plungers inserted into rotor body recesses; *D*, elevator; *E*, elevator guide; *F*, control unit. *Bottom*: *A*, plunger rods and plungers; *B*, plunger inserts for rotor body recesses; *C* rubber caps; *D*, gradient–sample feed devices; *E*, sealing caps. (Constructed by and available from H. R. White Co., Simi Valley, Ca.)

ters the distributor as a single stream from above and may be linear, exponential, or take any other desired shape. A single sample may then be apportioned simultaneously onto the light end of the gradients, or different samples may be separately added. In the apparatus shown in Fig. 7-8, the piston that raises and lowers the tube plungers is moved hydraulically. After the tubes are loaded, they are removed from the apparatus for centrifugation using either swinging-bucket, fixed-angle, or vertical tube rotors.

Direct extension of this concept for parallel gradient formation to vertical tube rotors is shown in Fig. 7-11. In this approach, the cavities of the rotor are *directly* filled in concert with the density gradient (and sample) by lowering plungerlike inserts housed in each of the rotor's cavities (Sheeler et al., 1980). Parallel vertical movement of the inserts is achieved by using the loading platform that contains a worm-gear mechanism and variable-speed, reversible motor. The annular array of openings in the floor of the rotor through which vertical displacement of the inserts is achieved is sealed when the inserts bottom out in each recess. After the gradients and the samples are loaded and the rotor removed from the loading platform, the feed devices mounted in the upper part of each recess are capped and sealed for centrifugation. On completion of centrifugation, the gradients and the entrained particle zones may be collected either in parallel or separately.

REFERENCES AND RELATED READING

Books

Birnie, G. D. and Rickwood, D. (Eds.), *Centrifugal Separations in Molecular and Cell Biology,* Butterworths, London, 1978.

Rickwood, D., *Metrizamide, A Gradient Medium for Centrifugation Studies.* Nyegaard, Oslo, Norway, 1978.

Rickwood, D. (Ed.) *Centrifugation: A Practical Approach.* Information Retrieval, London, 1978.

Articles and Reviews

Brunk, C. F. and Leick, V. (1969), Rapid equilibrium isopycnic CsCl gradients. *Biochim. Biophys. Acta* **179,** 136.

Candler, E. L., Nunley, C. E., and Anderson, N. G. (1967), Analytical techniques for cell separations. VI. Multiple gradient-distributing rotor (B-XXI). *Anal. Biochem.* **21,** 253.

Flamm, W. G., Birnstiel, M. L., and Walker, P. M. B. (1969), Preparation and fractionation, and isolation of single strands of DNA by isopycnic ultracentrifugation in fixed-angle rotors, in *Subcellular Components* (G. D. Birnie and S. M. Fox, Eds.), Butterworths, London.

Gregor, H. D. (1977), A new method for the rapid separation of cell organelles. *Anal. Biochem.*, **82**, 255.

Hell, A., Rickwood, D., and Birnie, G. D. (1974), Buoyant density-gradient centrifugation in solutions of metrizamide, in *Methodological Developments in Biochemistry*, Vol. 4, *Subcellular Studies* (E. Reid, Ed.), Longmans, London.

Hofman, L. F., Moline, C., McGrath, G., and Barron, E. J. (1978), Use of vertical rotors to facilitate estrogen and progesterone receptor analysis. *Clin. Chem.* **24**, 1609.

Jordon, V. C., and Prestwich, G. (1977), Binding of (^{3}H)-tamoxifen in rat uterine cytosols: A comparison of swinging-bucket and vertical tube rotor sucrose density gradient analysis. *Molec. Cell Endocrinol.* **8**, 179.

Masket, A. V. (1941), A quantity type rotor for the ultracentrifuge. *Rev. Sci. Instrum.* **12**, 277.

Pertoft, H. and Laurent, T. C. (1969), The use of gradients of colloidal silica for the separation of cells and subcellular components, in *Modern Separation Methods of Macromolecules and Particles* (T. Gerritsen Ed.), Wiley-Interscience, New York.

Pertoft, H., Laurent, T. C., Laas, T., and Kagedal, L. (1978), Density gradients prepared from colloidal silica particles coated by polyvinylpyrrolidone (Percoll). *Anal. Biochem.* **88**, 271.

Sheeler, P. and Doolittle, M. H. (1979), unpublished observations.

Sheeler, P., Doolittle, M. H., and White, H. R. (1978), Method and apparatus for producing and collecting a multiplicity of density gradients. *Anal. Biochem.* **87**, 612.

Sheeler, P., White, H. R., and Herzog, N. (1980), Parallel production of density gradients in a modified vertical tube rotor (in preparation).

Siaskotos, A. N. and Wirth, M. E. (1967), A method for the mass production of density gradients. *Anal. Biochem.* **19**, 201.

Wells, J. R. and Brunk, C. F. (1979), Rapid CsCl gradients using a vertical rotor. *Anal. Biochem.* **97**, 196.

EIGHT

Influence of Rotor Geometry on Density Gradient Profiles and Resolution

SPECIAL-PURPOSE DENSITY GRADIENTS

It should be clear from the preceding chapters that one can make a choice among several fundamentally different types of rotor when planning centrifugal separations of particles in density gradients. Although the swinging-bucket rotor remains the workhorse in this area, density gradient separations are also regularly carried out in fixed-angle, zonal, and vertical tube rotors. Most of the commonly encountered density gradient makers produce gradients in which there is a specific and predictable mathematical relationship between gradient volume and gradient density (or, to be more precise, gradient *concentration*); accordingly, there are "linear," "concave exponential," "convex exponential," and other gradient shapes.

For the most part, except at the hemispherical or conical end of a centrifuge tube, the relationship in a swinging-bucket rotor between *radius* and density is the same as that between *volume* and density. For example, if a gradient in which the density varies linearly with volume as it emerges from the gradient maker is used to fill a centrifuge tube, the gradient will also vary linearly with radius during centrifugation in a swinging-bucket rotor (Fig. 8-1). The volume–density profiles of other types of gradients will also be retained in the radial coordinates of a horizontally positioned centrifuge tube. A similar although slightly altered correspondence exists in fixed-angle rotors (see below). In zonal rotors, however, there is a

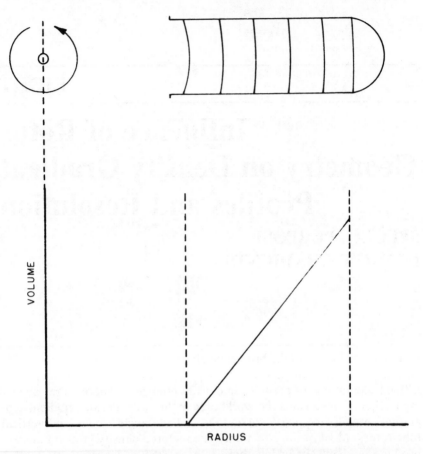

Fig. 8-1 Except in the hemispherical end of the centrifuge tube, a linear relationship between radius and volume exists in a swinging-bucket rotor.

significant departure in the volume–density–radius relationship, and this is due to the sector shape of the rotor's compartments. As we shall see, sector shape also affects the resolution obtainable with the use of specific density gradients. Comparisons between separations in tubes and in zonal rotors and the design of experiments in one rotor format based on the results obtained in another format must take into account these differences in geometry. In this chapter we consider a number of different special-purpose density gradients and reconcile them with the specific geometry encountered in the various rotors used for particle separations.

LINEAR AND ISOMETRIC GRADIENTS

Unfortunately, the expression "linear gradient" can take on several different meanings. In its earliest and simplest form, the expression referred to a gradient in which the density (or solute concentration) changed in direct proportion to the volume of gradient produced. This is the type of gradient generated by the simple, two-chamber devices described in Chapter 4. In a centrifuge tube that has been filled with such a gradient and placed in a swinging-bucket rotor, the density also varies directly with radius once the bucket has swung into the horizontal position. However, when the same gradient fills a spinning zonal rotor, the density does *not* vary linearly with radius; instead, the gradient density varies in a concave exponential manner (Fig. 8-2). This is due to the fact that volume increases exponentially with radius in the sector shape chambers of a zonal rotor.

If we discount the space taken up by the septa and also consider the rotor bowl floor and lid to be parallel, then the volume found between the edge of the core and any radial position in the rotor bowl is given by the relationship

$$V = \pi h (r_x^2 - r_c^2) \tag{8-1}$$

where h is the internal height of the rotor, r_x is the distance from the axis of rotation to any selected radial position, and r_c is the radius of the core. For the density of a gradient placed in a zonal rotor to increase linearly with radius, its density must increase in a convex exponential fashion with respect to volume. Such a gradient could be produced by an exponential gradient maker in which the ratio of the volume of the mixing chamber to the total delivered volume is selected on the basis of the h, r_c, and r_x values of the zonal rotor. To avoid the confusion surrounding the use of the term "linear," gradients in which density varies in direct proportion to radius are now called "isometric gradients." Hence an isometric gradient for a swinging-bucket rotor could be made by using a simple linear gradient maker, whereas an isometric gradient for a zonal rotor requires either an exponential gradient-making device or other special equipment.

ISOKINETIC GRADIENTS

Isokinetic gradients, proposed first for conventional tube-format centrifugation by Noll (1967), are density gradients within which a given family

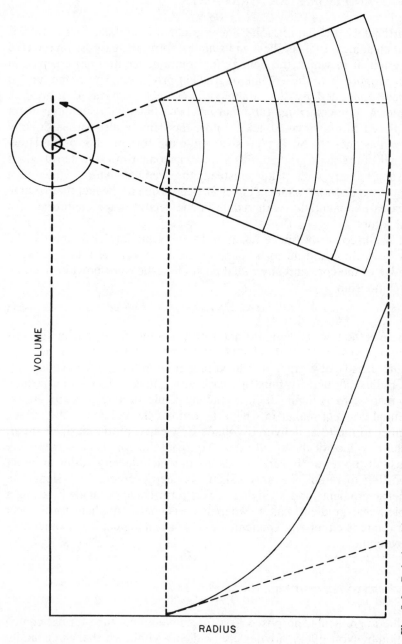

VOLUME

RADIUS

Fig. 8-2 Relationship between radius and volume in a zonal rotor. (For comparison, the relationship in a centrifuge tube is also shown; i.e., the dashed line.)

of particles sediments at a constant speed. In such a gradient, the increasing density, viscous drag and the buoyancy exactly compensate for the increasing centrifugal force experienced by the particles as they move further away from the axis of rotation. To achieve this, the ratio

$$\frac{(\rho_P - \rho_M)}{\eta} x \tag{8-2}$$

must be constant. Substituting a constant, K_1, for the ratio in 8-2 into equation 4-2 we obtain

$$\frac{dx}{dt} = \frac{2r^2 (\rho_P - \rho_M)}{9\eta} \omega^2 x = (\tfrac{2}{9})r^2 K_1 \omega^2 = K_2 \tag{8-3}$$

Consequently, for a given population of particles in which r (particle radius) remains constant and rotor speed (measured as ω) is also kept constant, dx/dt will remain constant (i.e., $dx/dt = K_2$, as in equation 8-3).

Particles that vary in size but have equal densities will sediment through an isokinetic gradient at speeds that are proportional to their $s_{20,w}$ values. Consequently, from an analytical point of view, isokinetic gradients can be used to estimate particle sedimentation coefficients or to anticipate the position of a particle zone in a density gradient after a given period of centrifugation. For preparative purposes, isokinetic gradients assist in the optimization of resolution by keeping particle zone widths constant. Since zone widths remain constant, the final thickness of a particle zone (and thus the resolution of different zones) is determined to a large degree by the sample zone thickness—a parameter that may readily be controlled.

The design of an isokinetic gradient must take into account (1) the specific rotor and tube being used in the experiment and (2) for the gradient solute being used, the precise relationship between concentration, density, and viscosity at the temperature of the rotor. For the gradient to be truly isokinetic, the sedimenting particles should be spherical and osmotically inert, although in practice slight deviations from spherical shape do not appreciably alter the anticipated results.

Instructions for the preparation of isokinetic gradients for particles of varying densities using a number of different Beckman Spinco and IEC swinging-bucket rotors may be found in the articles of Noll (1967) and McCarty et al. (1968) and also in the *Handbook of Microbiology*. Steensgaard (1970) and Steensgaard and Hill (1970) have described isokinetic gradients for the B-XIV and B-XV zonal rotors. The profiles for published isokinetic gradients were initially determined by computer solution of equations that took into account the variables described above. Unfortunately, these solutions have been obtained for sucrose gradients, and not for other gradient solutes. Equally effective isokinetic gradient profiles can be determined without computer approximation by calculating the concentration of gradient solute necessary to keep expression 8-2 constant for several values of x and then fitting a smooth curve through the resulting points. Since the relationship between ρ_M and η is known for a number of other gradient solutes (see the Appendixes), isokinetic gradient separations are not restricted to sucrose.

Isokinetic gradients characteristically assume a gently convex exponential shape, and for this reason they can be closely approximated by using a simple exponential gradient maker if the limiting concentrations and the volumes of the reservoir and mixing chambers are properly selected.

SECTORIAL DILUTION IN ZONAL ROTORS

In density gradient centrifugation, as in other zone-separation techniques such as electrophoresis and chromatography, a major goal is to restrict the widths of zones of monodisperse particles (or molecules) moving away from their starting positions (i.e., moving away from the sample zone). In electrophoresis and chromatography, one of the principal causes of zone broadening (with the resulting loss of resolution) is *diffusion*. However, during the centrifugal fractionation of cells or subcellular particles the effects of diffusion are only minor. Accordingly, as noted in the preceding discussion, isokinetic gradients provide an effective means for controlling zone widths. In zonal rotors, an additional problem is created by the sector-shaped compartments in that resolution may be lost even in isokinetic gradients. The culprit is *sectorial dilution*.

In a zonal rotor the volume V occupied by a cylindrical zone is given by

$$V = \pi h (r_2^2 - r_1^2) \qquad (8\text{-}4)$$

where h is the height of the zone and r_1 and r_2 are its respective minimum and maximum radii. For simplicity, the space occupied by the septa has been discounted. It may be seen from equation 8-4 that even though the zone thickness may remain constant (i.e., $r_2 - r_1$ is constant), V increases as r_2 and r_1 increase. For example, in a rotor of internal height 5 cm, a 1-cm-thick zone extending from 3 to 4 cm from the axis of rotation occupies 110 ml, whereas the same thickness zone extending from 8 to 9 cm would have a volume of 267 ml. Consequently, the concentration of 10,000 particles in this zone would drop from 91 to 37 particles per milliliter as the zone sedimented radially, *even though the zone thickness remained the same.*

Steensgaard and Hill (1970) were the first to adapt the isokinetic approach to zonal rotors and to design density gradients in which sectorial expansion could be taken into account to the extent that particle zones remained uniformly thick. To produce isokinetic gradients in zonal rotors, the precise relationship between x (i.e., rotor radius) and volume must be known, a relationship that varies from one zonal rotor to another. Although useful from an analytical point of view, isokinetic density gradients do not offer the same advantages in zonal rotors as in centrifuge tubes. To achieve improved resolution during preparative fractionation of particle mixtures, the density gradient must be designed to counterbalance the sectorial effect.

EQUIVOLUMETRIC GRADIENTS

For density gradient centrifugation in centrifuge tubes of conventional shape, particles sedimenting through an isokinetic gradient pass through equal volumes of gradient in equal increments of time. As we have noted, this is not true for zonal rotors where particles in zones of constant thickness sedimenting isokinetically occupy ever-increasing volumes. To improve resolution in zonal rotors, particles at the leading edge of a zone must always be sedimenting at a rate less than that of equivalent particles at the rear edge of the zone. In this way, zone thickness continually diminishes. A gradient in which the decrease in zone width during sedimentation exactly counterbalances the sectorial effect so that the *volume* occupied by the particle zone remains constant is known as an *equivolumetric* gradient (Pollack and Price, 1971). For density gradient centrifugation in the tubes of a swinging-bucket rotor, isokinetic gradients are simultaneously equivolumetric. In zonal rotors, for the volume of a cylindrical zone to remain constant as its distance from the axis of rotation increases, the density gradient would have to satisfy the following requirements:

$$x \left(\frac{\rho_P - \rho_M}{\eta} \right) x = \frac{x^2 (\rho_P - \rho_M)}{\eta} = K \tag{8-5}$$

The equivolumetric density gradient of the zonal rotor may be thought of as the sectorial equivalent of the isokinetic gradient. Equivolumetric gradients for B-XIV, B-XV, B-XXIX, and B-XXX zonal rotors are described by Pollack and Price (1971), Price and Hsu (1971), van der Zeijst and Bult (1972), Price (1973), and Eikenberry (1973). The value of the equivolumetric density gradient lies in the fact that the volumetric distance moved by particles during centrifugation is proportional to their sedimentation coefficients, and the widths of particle zones are independent of the radial distance migrated. The profiles of equivolumetric, isokinetic, and isometric density gradients in the TZ-28 reorienting gradient zonal rotor are compared in Fig. 8-3.

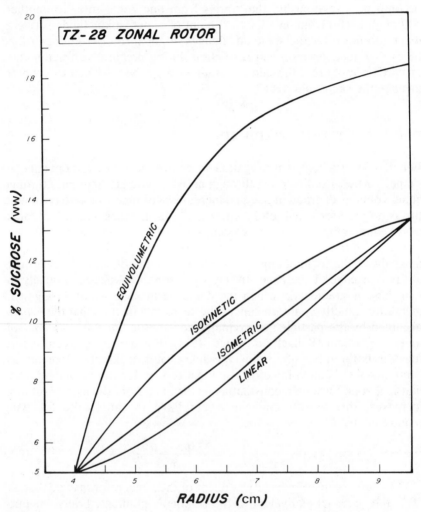

Fig. 8-3 A comparison of the radius–solute concentration profiles for linear, isometric, isokinetic, and equivolumetric sucrose gradients in the TZ-28 zonal rotor.

GRADIENT PROFILES AND FIXED-ANGLE AND VERTICAL TUBE ROTORS

So far we have compared the relationship between centrifuge tube or rotor radius and gradient shape for swinging-bucket and zonal rotors. We now turn to a similar discussion of fixed-angle and vertical tube rotors. Since fixed-angle and vertical tube rotors are used principally for isopycnic separations of particles and rarely for rate separations, our discussion can be greatly simplified, for in an isopycnic separation the final thickness of a particle zone is more or less independent of the sample thickness or the specific shape of the density gradient (see Chapter 5).

Fixed-Angle Rotors The specific phenomena that characterize particle sedimentation through density gradients in fixed-angle rotors were considered in some depth in Chapter 5 and are not repeated here. Insofar as the density gradients in such rotors are concerned, once reorientation has occurred, the sedimenting particles pass through a continuum of elliptical isodense planes until either centrifugation is terminated or the particles form elliptical bands at their isopycnic positions. The specific angle of inclination maintained when a centrifuge tube containing gradient and sample is placed in a fixed-angle rotor results in a radius–density profile that is different from that observed if the tube were placed in a swinging-bucket rotor. This is depicted in Fig. 8-4 using the simplest example, namely, an isometric (or "linear") gradient. The diminished radial length of a density gradient in a fixed angle rotor means that the sedimenting particles "see" a *steeper* gradient than in a swinging bucket rotor. Therefore, at equivalent radii, particle sedimentation rates are lower in the fixed-angle rotor. Results obtained under specified conditions in swinging-bucket rotors cannot be directly extrapolated to fixed-angle rotors.

Except at the hemispherical end of the centrifuge tube (which generally may be ignored), vertical sections through a tube positioned in a fixed-angle rotor have equal, elliptical areas. Consequently, linear and isometric gradients are the same in fixed-angle rotors, and so too are isokinetic and equivolumetric gradients. Again, it should be noted that these special gradients are most often used in conjunction with the analysis of particle sedimentation rates or the resolution of different particles on the basis of differences in their s values. Such studies would normally be carried out in swinging-bucket or in zonal rotors and only rarely attempted in fixed-angle rotors.

Vertical Tube Rotors The use of vertical tube rotors for density gradient separations was considered in Chapter 7. Since wall effects are not as marked in vertical tube rotors as they are in fixed-angle rotors, vertical tube rotors can be used for rate separations as well as for isopycnic banding. The cross-sectional surface through which particles sediment once reorientation has occurred may be thought of as tall rectangles whose widths increase in the first half of the tube and which decrease in the second half (Fig. 8-5). Consequently, a gradient whose density varies linearly with volume takes on a sigmoid shape when considered as a function of radius (Fig. 8-5). Since the total radial length of a density gradient in a vertical tube rotor is the diameter of the centrifuge tube, a given

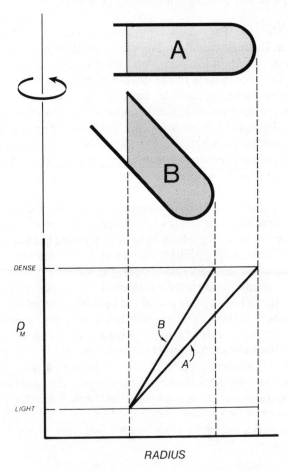

Fig. 8-4 The radial length of a density gradient in a fixed-angle rotor is less than in a swinging-bucket rotor; consequently, the gradient is steeper.

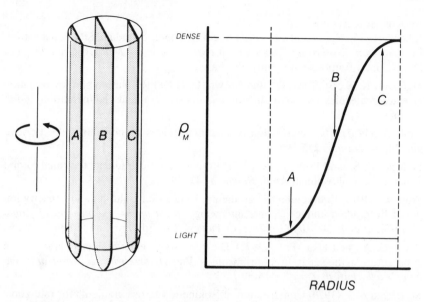

Fig. 8-5 The areas of the isodense planes through which particles sediment in a vertical tube rotor increase in the first half of the tube's diameter and decrease in the second half. Gradients in which density varies linearly with volume thus assume a sigmoid density–radius profile.

gradient would be steepest in a vertical tube rotor, less steep in a fixed-angle rotor, and least steep in a swinging-bucket rotor.

The differences in shape, cross-sectional area, sedimentation path length, and other properties that distinguish vertical tube rotors, fixed-angle rotors, and swinging-bucket rotors affect not only particle sedimentation rates in density gradients, but also the relative efficiency of these rotors during differential pelleting. These differences are considered in Chapter nine.

REFERENCES AND RELATED READING

Books

Birnie, G. D., and Rickwood, D. (Eds.), *Centrifugal Separations in Molecular and Cell Biology*. Butterworths, London, 1978.

Laskin, A. I., and Lechevalier, H. A. (Eds.), *Handbook of Microbiology*, Vol. II, *Microbial Composition*. CRC Press, Cleveland, 1973.

Hinton, R., and Dobrota, M. *Density Gradient Centrifugation*. North-Holland, Amsterdam, 1976.

Articles and Reviews

Eikenberry, E. F. (1973) Generation of equivolumetric gradients for zonal rotors. In *European Symposium for Zonal Centrifugation in Density Gradient*, (G. Dorvyl, Ed.) Editions Cite Nouvelle, Paris.

McCarty, K. S., Stafford, D., and Brown, O. (1968) Resolution and fractionation of macromolecules by isokinetic sucrose density gradient sedimentation. *Anal. Biochem.*, **24**, 314.

Noll, H. (1967) Characterization of macromolecules by constant velocity sedimentation. *Nature*, **215**, 360.

Pollack, M. S., and Price, C. A. (1971) Equivolumetric gradients for zonal rotors: Separation of ribosomes. *Anal. Biochem.* **42**, 38.

Price, C. (1973) Equivolumetric gradients: Limits on resolution and capacity imposed by gradient-induced zone narrowing. In *European Symposium of Zonal Centrifugation in Density Gradient*, G. Dorvyl, Ed. Editions Cite Nouvelle, Paris.

Price, C. A., and Hsu, T. S. (1971) The capacity of equivolumetric gradients in zonal rotors in the separation of ribosomes. *Vierteljahrschrift der Naturforschenden Gesellschaft in Zurich*, **116**, 367.

Steensgaard, J. (1970) Construction of isokinetic sucrose gradients for rate-zonal centrifugation. *Eur. J. Biochem.*, **16**, 66.

Steensgaard, J., and Hill, R. J. (1970) Separation and analysis of soluble immune complexes by rate zonal ultracentrifugation. *Anal. Biochem.*, **34**, 485.

van der Zeijst, A. M., and Bult, H. (1972) Equivolumetric glycerol and sucrose gradients for the B-XV zonal rotors. *Eur. J. Biochem.*, **28**, 463.

NINE

Predicting Run Conditions

It is not unusual for many investigators, especially those to whom centrifugation is either a new or an occasional technique, to determine run conditions by "trial and error" or to follow step by step a procedure described in the literature for a comparable particle separation. In the latter case the investigator may seek to use precisely the same rotor, centrifuge, run time, speed, temperature, and so on. Although these approaches may provide the desired results after one or more attempts, it is also possible to closely approximate the run conditions necessary to affect a particular separation by making a few simple mathematical calculations based on some of the expected properties of the particles to be isolated and the known specifications of the rotors and centrifuges available to the laboratory. The latter approach is frequently more efficient, providing run conditions that are tailored to the equipment being used and that may also be considerably simpler than otherwise attempted. In this chapter we examine some ways that one can predict run conditions for both differential pelleting of particles and for zone separations in density gradients.

ROTOR EFFICIENCY—THE K FACTOR

The differential equation defining the sedimentation coefficient

$$s = \frac{dx/dt}{\omega^2 x} \tag{9-1}$$

may be integrated to yield an algebraic relationship as follows: If a boundary is x_1 centimeters from the axis of rotation at time t_1 and x_2 centimeters at time t_2, then equation 9-1 may be solved by integration. First, by transposition,

$$s \, dt = \frac{1}{\omega^2} \frac{dx}{x} \tag{9-2}$$

integrating between the limits set above, we obtain

$$s \int_{t_1}^{t_2} dt = \frac{1}{\omega^2} \int_{x_1}^{x_2} \frac{dx}{x} \tag{9-3}$$

and

$$s(t_2 - t_1) = \frac{1}{\omega^2} (\ln x_2 - \ln x_1) = \frac{1}{\omega^2} \left(\ln \frac{x_2}{x_1} \right) \tag{9-4}$$

Therefore,

$$s = \frac{1}{\omega^2(t_2 - t_1)} \ln \frac{x_2}{x_1} \tag{9-5}$$

where t is expressed in seconds.

Now, if x_2 and x_1 respectively represent the maximum and the minimum radii of the rotor chambers (e.g., R_{max} and R_{min}), equation 9-5 becomes

$$s = \frac{\ln(R_{max}/R_{min})}{\omega^2(t_2 - t_1)} \tag{9-6}$$

A helpful expression used in centrifugation is the K (or "clearing") factor, which helps to predict the time it will take to move a boundary of particles from the centripetal edge of the rotor compartment (i.e., R_{min}) to the centrifugal edge (i.e., R_{max}). K is defined as follows:

$$K = (T)(s)10^{13} \quad \text{or} \quad (T)(S) \tag{9-7}$$

In this equation, T is the time required (in hours) and S is the sedimentation coefficient (in Svedberg units).

Substituting the definition of s in equation 9-6 for s in equation 9-7 yields

$$K = \frac{T \ln(R_{max}/R_{min})10^{13}}{\omega^2(t_2 - t_1)} \tag{9-8}$$

Since T (in hours) is equal to $(t_2 - t_1)/3600$ (i.e., there are 3600 sec/hr), equation 9-8 becomes

$$K = \frac{\ln(R_{max}/R_{min})10^{13}}{3600\omega^2} \tag{9-9}$$

The angular velocity and the speed are related as follows

$$\omega = \left(\frac{\text{rpm}}{60}\right)(2\pi)$$

Substitution of this value of ω into equation 9-9, followed by simplification, yields

$$K = \frac{\ln(R_{max}/R_{min})10^{13}}{3600[(\text{rpm}/60)(2\pi)]^2} = \frac{\ln(R_{max}/R_{min})10^{13}}{(\text{rpm})^2 39.5} \qquad (9\text{-}10)$$

Thus

$$K = \frac{2.53 \times 10^{11}\ln(R_{max}/R_{min})}{(\text{rpm})^2} \qquad (9\text{-}11)$$

Equation 9-11 may now be conveniently used to calculate the clearing factor of any rotor of known cavity dimensions operated at a specified speed.

The lower the K factor, the more efficient the rotor in sedimenting particles. The K factors for most commercial rotors are listed in the Appendixes. If the K factor for a rotor operated at its maximum speed is known, then the corresponding K factor at a lower speed is obtained simply by multiplying by $(\text{rpm}_{max}/\text{rpm})^2$.

Example 1 The DuPont/Sorvall AH-627 swinging-bucket rotor has a maximum speed of 27,000 rpm, an R_{max} of 16.6 cm, and an R_{min} of 7.7 cm. The K factor at the maximum operating speed would be

$$\frac{2.53 \times 10^{11}\ln(16.6/7.7)}{(2.7 \times 10^4)^2} = 267$$

Note that 267 is the K factor at 27,000 rpm; at any speed below this the K factor would be higher. For example, at 20,000 rpm the rotor's K factor would be 267 $(27,000/20,000)^2 = 486$.

Example 2 The Beckman Spinco Type 30 fixed-angle rotor has a maximum speed of 30,000 rpm, an R_{max} of 10.5 cm, and an R_{min} of 5.0 cm. The K factor at the maximum operating speed would be

$$\frac{2.53 \times 10^{11}\ln(10.5/5.0)}{(3.0 \times 10^4)^2} = 209$$

Example 3 The DuPont/Sorvall SV-288 vertical tube rotor has a maximum operating speed of 20,000 rpm, an R_{max} of 9.02 cm, and an R_{min} of

6.47 cm. The K factor at the maximum operating speed would be

$$\frac{2.53 \times 10^{11} \ln(9.02/6.47)}{(2 \times 10^4)^2} = 210$$

Predicting Run Times   One of the most important contributions of the knowledge of the K factor is that it may be used in equation 9-7 to estimate the time required to clear (i.e., pellet) the particles present in suspension by centrifugation in a given rotor at a given speed.

Example 1   For 100S particles in a Beckman Type 30 rotor at 30,000 rpm (i.e., K factor 209), the time required would be $T = K/S = 209/100 = 2.09$ hr.

Example 2   For 200S particles in a DuPont/Sorvall AH-627 rotor at 27,000 rpm (i.e., K factor 267), the time required to pellet the particles would be $T = K/S = 267/200 = 1.34$ hr.

And so on. It should be noted that the T value is based on particle s values. Therefore, clearing times will vary if the particles are not suspended in pure water at 20°C.

An examination of the K factors of a variety of rotors reveals an important generalization. Vertical rotors are more efficient than fixed-angle rotors, and these, in turn, are more efficient than swinging-bucket rotors. The principal reason is discerned from an examination of Fig. 7-6, which shows how the absolute values of R_{max} and R_{min} as well as their relative values differ among the three rotor types. Generally speaking, the maximum sedimentation path (i.e., R_{max} minus R_{min}) is lowest for vertical tube rotors, whereas R_{min} is lowest in swinging-bucket rotors, higher in fixed-angle rotors, and highest in vertical tube rotors. It should also be noted that the inherent simplicity of vertical tube rotors and fixed-angle rotors (e.g., no moving parts) and their greater natural stability and more efficient aerodynamic shape provide for higher operating speeds while accepting the same or greater quantities of sample (see Appendixes).

In Chapter 3, it was noted that the wall effects that characterize particle sedimentation in fixed-angle rotors materially improve their effectiveness during pelleting or isopycnic banding of particles. These influences are not taken into account in using K factors to estimate run times. Therefore, it should be expected that the minimum run time necessary to form a pellet using a fixed-angle rotor may actually be less than that predicted by equation 9-7.

PREDICTING RUN TIMES IN DENSITY GRADIENTS

In the preceding section it was shown that the speed and the time necessary for pelleting of a suspension of particles may be closely approximated if one knows certain rotor parameters as well as the sedimentation coefficient of the particles. It should be emphasized that the equations derived earlier apply to particles suspended in pure water at 20°C. If the suspending liquid is not pure water or the temperature is not 20°C, the sedimentation coefficient must be adjusted to account for the liquid's altered density and viscosity. The following relationship may be used to estimate the particles' sedimentation coefficient under the experimental conditions actually being used:

$$s_{T,m} = s_{20,w} \left[\frac{\eta_{20,w}(\rho_P - \rho_{T,m})}{\eta_{T,m}(\rho_P - \rho_{20,w})} \right] \tag{9-12}$$

In this equation $s_{T,m}$ is the sedimentation coefficient in the suspending liquid being used, $s_{20,w}$ is the sedimentation coefficient in pure water at 20°C, $\eta_{20,w}$ and $\rho_{20,w}$ are respectively the viscosity and the density of water at 20°C, $\eta_{T,m}$ and $\rho_{T,m}$ are the viscosity and the density of the suspending liquid at the temperature of the run, and ρ_P is the density of the particles.

In the majority of instances, the differential pelleting of suspended particles is carried out in buffered saline solutions. In these instances $s_{20,w}$ and $s_{T,m}$ do not differ greatly, and run conditions can be estimated simply by using the $s_{20,w}$ value. However, in the case of density gradient centrifugation, (1) $\rho_{T,m}$ and $\eta_{T,m}$ even at the light end of the gradient differ considerably from $\rho_{20,w}$ and $\eta_{20,w}$, and (2) $\rho_{T,m}$ and $\eta_{T,m}$ undergo *continuous change;* consequently, the estimation of run times becomes considerably more complex.

One possible mathematical solution to this problem is to divide the density gradient being used into a series of small radial increments (Δx) and determine the average density and viscosity of the gradient in each increment (i.e., $\bar{\rho}_M$ and $\bar{\eta}_M$). If this is done, run conditions, expressed as $\omega^2 t$ (see below), may be approximated from the relationship

$$\omega^2 t = \frac{\rho_P - \rho_{20,w}}{s_{20,w}\eta_{20,w}} \sum_{x_1}^{x_2} \frac{\bar{\eta}_M \Delta x}{(\rho_P - \bar{\rho}_M)x} \tag{9-13}$$

where $\bar{\eta}_M$ and $\bar{\rho}_M$ are the respective viscosity and the density of the gradient in the center of each radial increment, x is the radial distance to the center of each increment, and x_1 and x_2 are the radial positions of the zone of particles at the beginning and at the end of the run. In general,

Table 9-1 K' Factors for Some Swinging-Bucket Rotors

Rotor	Maximum rpm	K' Factor, Particle Density			
		1.2	1.4	1.6	1.8
Beckman Spinco					
SW 65 Ti	65,000	143	120	113	110
SW 60 Ti	60,000	160	135	128	124
SW 41 Ti	41,000	379	317	300	292
SW 25.1	25,000	1,035	867	822	799
DuPont/Sorvall					
AH 650	50,000	184	153	145	141
AH 627	27,000	816	683	646	629

the use of equation 9-13 is far too tedious to be practical, and other methods are to be preferred.

The K' Factor   The K' factor is a rotor-specific factor that can be used to estimate the amount of time necessary to sediment particles of known density and $s_{20,w}$ value through a linear (i.e., isometric) 5 to 20% w/w sucrose density gradient at 5°C. The K' factors for various rotors can usually be obtained from their manufacturers, and a number of examples are given in Table 9-1. Like the K factor discussed earlier, the lower the K' factor, the more efficient the rotor in sedimenting particles. The K' factor is used to estimate run times by employing the following equation:

$$K' = (T)(s)10^{13} \quad \text{or} \quad (T)(S) \tag{9-14}$$

In this equation T is the time required (in hours), and S is the sedimentation coefficient in Svedberg units.

Example   How long would it take to sediment 80S particles of density 1.4 through a 5 to 20% w/w sucrose density gradient at 5°C using the Beckman Spinco SW 65Ti rotor at its maximum speed?
   From Table 9-1, we know that K' is 120 at 65,000 rpm; therefore,

$$T = \frac{120}{80} = 1.5 \text{ hr}$$

The K' factor is a very convenient expression, but its value is limited since it applies to a single type of density gradient (i.e., 5 to 20% w/w sucrose at 5°C). Equally convenient but much more useful are aids to sedimentation estimation available from the various centrifuge companies that take the form of graphs and nomograms.

$s\omega^2 t$ **Charts**   Especially useful for predicting run times are the $s\omega^2 t$ charts produced by Beckman Instruments. The charts are designed for use with either 5 to 20% w/w or 15 to 30% w/w isometric sucrose density gradients at temperatures of 4°C or 20°C. Separate charts are available for each of their swinging-bucket and zonal rotors. With the charts, the time required to sediment particles of any known ρ_P and $s_{20,w}$ value any distance through the gradient at any selected rotor speed can be quickly estimated.

One of the $s\omega^2 t$ charts for the SW 50.1 rotor is reproduced in Fig. 9-1. The following conditions were met in preparing the charts (and must be satisfied if they are to be used properly): (1) the sucrose density gradient is isometric (in this example, linear with respect to volume); (2) no cushion is used (3) the starting position of the sample is 3 mm below the top of the tube; and (4) the maximum radius is taken to be the end of a cylinder that has the same volume as the tube. (Since the bottom of the centrifuge tube is hemispherical, the maximum cylinder radius is slightly less than that of the corresponding centrifuge tube.) Several additional conditions prevail in the $s\omega^2 t$ charts for Beckman zonal rotors. Although each chart contains curves only for particles of density 1.4 and 1.8, interpolation around the curves for particles of other densities will yield satisfactory results.

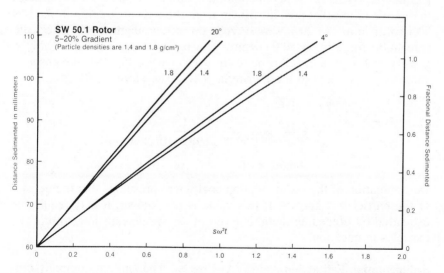

Fig. 9-1 $s\omega^2 t$ Chart for the SW 50.1 rotor. (Courtesy of Beckman Instruments Inc.).

The charts are used in the following way (refer to Fig. 9-1). Having selected the desired rotor and speed, choose the distance that you wish the particles to travel through the gradient and draw a horizontal line from this position on the ordinate axis to the appropriate curve. From the point of intersection with the curve, draw a perpendicular line down to the abscissa and read the value for $s\omega^2 t$. If the s value of the particles being sedimented in known, t (in seconds) may readily be calculated.

Example 1 How long would it be necessary to centrifuge particles that have a sedimentation coefficient of $40S$ and a density of 1.4 in the SW 50.1 rotor at 30,000 rpm and at 4°C in a 5 to 20% w/w sucrose gradient to move the particles to a radial position 8.7 cm from the axis of rotation? Referring to Fig. 9-1 and following the instructions for using the $s\omega^2 t$ chart that were given above, the resulting $s\omega^2 t$ value would be 0.88. Therefore,

$$s\omega^2 t = 0.88$$

$$t = \frac{0.88}{s\omega^2} = \frac{0.88}{(40 \times 10^{-13})\,[(30{,}000/60)(2\pi)]^2}$$

$$t = 2.23 \times 10^4 \text{ sec or } 6.19 \text{ hr (6 hr, 11 min)}$$

Of course, the $s\omega^2 t$ charts can also be used to estimate the s value of an unknown particle.

Example 2 What is the sedimentation coefficient of a family of particles that move to a radial position 10 cm from the axis of rotation in the SW 50.1 rotor in a 5 to 20% w/w sucrose density gradient at 4°C after centrifugation for 2 hr at 50,000 rpm? The particles are known to have a density of 1.4. Again, referring to the $s\omega^2 t$ chart in Fig. 9-1, the abscissa value corresponding to these criteria is 1.3. Therefore,

$$s\omega^2 t = 1.3$$

$$s = \frac{1.3}{\omega^2 t} = \frac{1.3}{[(50{,}000/60)(2\pi)]^2(2)(60)(60)}$$

$$s = 6.59 \times 10^{-12} \text{ sec} \quad \text{or} \quad 65.9S$$

Determination of the sedimentation coefficient presumes that the density of the particles is known. If this value is not known, it may readily be estimated by preceding the above run by *isopycnic* centrifugation of the particles in question.

Sedimentation Estimation Aid (SEA) Charts   The DuPont company provides SEA charts with many of their rotors, and these, too, provide invaluable assistance in predicting run times for particles of known s and

ρ_P values or for estimating sedimentation coefficients experimentally. Sedimentation estimation aid charts are available for several different kinds of linear gradients at 4°C, including 5 to 10%, 5 to 20%, 5 to 30%, and 5 to 40% w/w sucrose. One of the SEA charts for the TV 850 vertical tube rotor is reproduced in Fig. 9-2.

To use the charts for predicting run times, one must know the s value and the density of the particles to be isolated. Alternatively, the charts may be used to determine the sedimentation coefficient of an unknown particle from its position in the gradient at the end of a run of known duration and speed. Let us consider both types of application of the charts by way of examples.

Example 3   The required volume of a 5 to 20% w/w sucrose gradient is prepared in tubes for the TV-850 rotor (in this instance, the required volume would be 32 ml; see Fig. 9-2), and a sample containing $10S$ particles of density 1.5 layered on top. What duration of centrifugation at 45,000 rpm would be required to sediment the particles to a position that is 15 ml from the top of the density gradient? Referring to Fig. 9-2, a horizontal line is first drawn from the 15-ml position on the right-hand ordinate axis until it intersects the 1.5 density curve; from this intersection a vertical line is drawn upward until it intersects the $10S$ diagonal. From this point, a horizonal line is drawn left until it meets the 45,000 rpm (i.e., 45 K rpm) diagonal. From this final intersection, a perpendicular line is then dropped to the abscissa to reveal the required time. In this example, the time required is 3.85 hr or 3 hr, 51 minutes.

Example 4   After centrifugation for 2 hr at 50,000 rpm (same density gradient, rotor, and so on, as in example 3 above), particles containing the particular biological activity under investigation are recovered in a fraction that is 24 ml from the top of the gradient. If the particles have been found to possess a density of 1.2, what is their sedimentation coefficient? First, draw a vertical line upward from the 2-hr position on the abscissa until it intersects the $50K$-rpm (i.e., 50,000-rpm) diagonal. From this intersection, draw a horizontal line to the right-hand ordinate axis. Now, plot a horizontal line from the 24-ml position on the right-hand ordinate until it intersects the 1.2 density curve. Plot a vertical line upward until it reaches the horizontal line drawn earlier. The intercept occurs on the $30S$ diagonal, thereby indicating that the sedimentation coefficient of the unknown particles is $30S$.

Using variations of the above procedures, one could estimate ρ_P (if s is known), determine the speed needed for a run of specific duration, and so on.

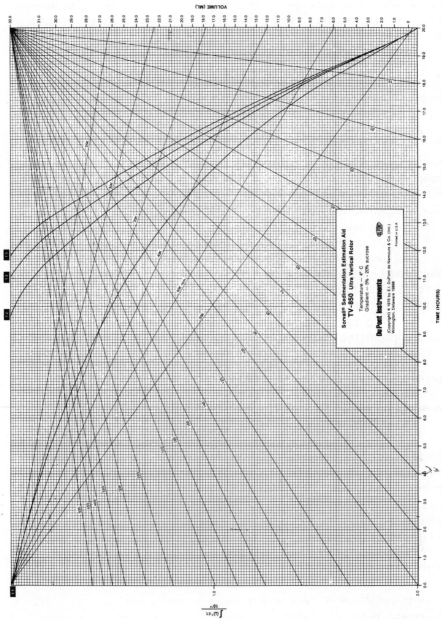

Fig. 9-2 Sedimentation estimation aid for the TV-850 vertical tube rotor. (Courtesy of E. I. DuPont and Company.)

178

$\omega^2 t$ **Integrators**   Most centrifuge companies offer an accessory for their centrifuges called an $\omega^2 t$ integrator (Fig. 9-3) that is especially useful in attacking problems of the sort discussed above. An $\omega^2 t$ integrator automatically keeps track of the accumulating centrifugal effect in the course of a run, *including* that taking place during acceleration and deceleration of the rotor.

If an entire run could be carried out at constant speed (i.e., if there were no period of acceleration or deceleration), the total centrifugal effect would simply be $\omega^2 t$ or

$$\left[\left(\frac{\text{rpm}}{60} \right) (2\pi) \right]^2 (t_2 - t_1)$$

where $t_2 - t_1$ is the time of speed in seconds. However, a determination of the additional centrifugal effect during acceleration of the rotor and during its deceleration requires the integration of changes in ω^2 over small time increments beginning from the moment that the rotor starts turning until the rotor finally comes to rest again; that is,

$$\int_{t_1}^{t_2} \omega^2 \, dt$$

An $\omega^2 t$ integrator provides this capability and also may be used to terminate a run so as to accumulate a particular $\omega^2 t$ value. The $\omega^2 t$ integrator is especially useful in two particular instances: (1) when it is used in

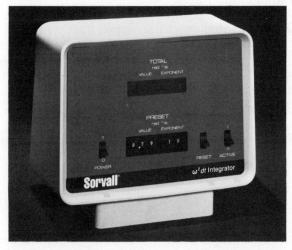

Fig. 9-3   An $\omega^2 t$ integrator. (Courtesy of E. I. DuPont and Company.)

conjunction with experiments with *dynamically unloaded zonal rotors,* the integrator allows one to keep track of the additional centrifugal force experienced by particles that continue to sediment during the unloading interval; and (2) when density gradients are to be reoriented in fixed-angle, vertical, and reograd zonal rotors, it is customary to extend the acceleration and deceleration time so that gradient (and particle) reorientation is both smooth and gradual. This added centrifugal effect during extended periods of rotor acceleration and deceleration is readily monitored and accounted for if an $\omega^2 t$ integrator is used.

An $\omega^2 t$ integrator can also be used in conjuction with SEA and $s\omega^2 t$ charts to simplify the determination or measurement of run conditions. In the case of Example 3 given earlier, by extending the horizontal line from its intersection with the 10S diagonal until it reaches the left-hand ordinate axis, one can read the required $\int \omega^2 \, dt$ needed for the run. This value (e.g., 3.1×10^{12} in the case of this particular example) may be set on the integrator, so that the run will automatically terminate at the proper time.

REFERENCES AND RELATED READING

Books
Birnie, G. D., and Rickwood, D., Eds. *Centrifugal Separations in Molecular and Cell Biology.* Butterworths, London, 1978.

Articles and Reviews
Bylund, D. B., and Bruening, G. (1974) Prediction of centrifugation time for equilibrium and velocity sedimentation on various gradients. *Anal. Biochem.,* **58,** 47.

McEwen, C. R. (1967), Tables for estimating sedimentation through linear concentration gradients of sucrose solution. *Anal. Biochem.,* **20,** 114. *Use of the $\omega^2 t$ Integrator. Applications Data DS-528.* Beckman Instruments Inc., Spinco Division, Palo Alto, California (1978).

TEN

Continuous-Flow Centrifugation

Continuous-flow centrifugation is an invaluable technique for efficiently isolating large amounts of particulate material from multiliter suspensions and is used extensively as a preparatory procedure in numerous biological, medical, and chemical studies. Probably the most common use of continuous-flow centrifugation is the harvesting of bacteria, algae, protozoa, or other cells grown in multiliter cultures as a preliminary to chemical, physiological, or morphological analysis. However, the technique is also frequently used to (1) collect cell- and particle-free culture media to isolate and assay cellular excretion products such as hormones, vitamins, enzymes, and growth substances; (2) separate on a continuous basis the cells and plasma of pooled, donated blood, (3) remove the larger subcellular components (such as nuclei and mitochondria) from large-volume tissue homogenates, (4) concentrate viruses from dilute suspensions, and (5) eliminate debris, precipitates, and other particulate contaminants from aqueous solutions.

During continuous-flow centrifugation, the suspension of particles is introduced as a continuous and uninterrupted stream into the centrifuge rotor *while it is spinning*. As the suspension passes through the rotor, particles are sedimented out of the stream and are trapped and concentrated within one or more rotor chambers, whereas the clarified liquid (i.e., the *supernatant*) exits the rotor and is collected separately. Continuous-flow rotors thereby eliminate the requirement of sequential *batch* separations when very large volumes of particle suspensions must be processed; consequently, these rotors can save considerable amounts of laboratory time and work. Once the entire suspension has been processed, the rotor is decelerated and opened, and the trapped particles are removed. The maximum rate at which suspensions can be processed by continuous-flow centrifugation (i.e., the maximum number of milliliters per minute) is determined by (1) the size, the shape, and the density of the particles to be collected and (2) the operating speed and the capacity of the rotor.

The particles present in very large volumes of suspensions can, of course, also be isolated by conventional methods, using swinging-bucket or fixed-angle rotors. However, this is usually far less efficient. Even the largest conventional swinging-bucket and fixed-angle rotors accommodate no more than about 6 l of suspension; therefore, a succession of spins is required when larger volumes of material must be handled. Furthermore, the increased size and mass of these rotors severely limits their maximum operating speeds and relative centrifugal forces, so that in most instances extended centrifugation time is required in order to ensure total particle "cleanout." Since continuous-flow rotors contain only a small percent of the total volume to be centrifuged at any instant in time, they may be quite small. Thus, in addition to eliminating the need for successive runs, continuous-flow rotors can be operated at much higher speeds; this implies higher RCF, lower K factors, and a more rapid and efficient particle cleanout.

Most continuous-flow rotors employ similar operating principles. Two channels communicate with the spinning rotor through a stationary apparatus that is mounted above and/or below the rotor. The particle suspension enters the rotor through one of these channels, and the supernatant exits the rotor through the other. The entry channel conducts the suspension into an axial opening in the spinning rotor where centrifugal force carries the stream radially toward the particle collection chambers. The supernatant is conducted from the rotor chambers back toward the rotor axis and into the exit channels. The openings through which liquid enters and leaves the collection chambers are arranged concentrically, with the entry ports positioned at a distance from the axis of rotation greater than that of the exit ports. As a result, liquid is caused to flow *centripetally* through each rotor chamber. The rate of centripetal flow is governed by the speed with which the suspension of particles is delivered to the spinning rotor.

The centrifugal force exerted on the particles as they enter the collection chambers is a function of the rotor speed and for each particle will define an instantaneous sedimentation rate. If this rate is *greater* than the rate at which the surrounding liquid is being displaced centripetally, the particles will sediment radially (i.e., centrifugally) and be trapped in the rotor compartments. However, if the sedimentation rate is *less* than the rate of centripetal flow, the particles are carried toward the exit ports of the compartments and out of the rotor. Usually, rotor speed and flow rate are selected to provide maximum cleanout of the particle suspensions. However, for heterogeneous populations of particles, the rotor speed and the flow rate can often be adjusted so that a differential *fractionation* of the particles is achieved. That is, depending on their sizes, shapes, and densities, some populations of particles will be trapped in the rotor while others are conducted out of the rotor with the supernatant.

Although the aims of all continuous-flow rotors are essentially the same, many different systems are available that offer a variety of specific applications. For example, some continuous-flow systems are "sealed;" that is, the particle suspension and the supernatant never contact the atmosphere and the rotor may be operated in the evacuated chamber of an ultracentrifuge. Such systems are especially useful in work with bio-hazardous materials and materials that readily form aerosols. Other continuous-flow systems are "open," where the particle suspension is fed across an air gap as it passes into the rotor. In some continuous-flow systems the trapped particles form a semisolid sediment (i.e., "pellet") at the centrifugal edge of the rotor's chambers, and in others the particles are continuously "banded" in a density gradient (see below). Particle collection chambers may take the form of centrifuge tubes, the sector-shaped compartments of a zonal rotor, a molded plastic liner inserted into the rotor, or a length of plastic tubing. In some continuous-flow instruments, the sediment as well as the supernatant is continuously expelled from the rotor.

For convenience, we divide continuous-flow formats into two major categories—*continuous flow with pelleting* (CFWP) and *continuous flow with banding* (CFWB)—and describe representative systems belonging to each of these categories.

CONTINUOUS FLOW WITH PELLETING

Collection in Centrifuge Tubes Probably the most popular continuous-flow system for collecting particles from suspension as a sediment in centrifuge tubes is the "KSB" (after its developers A. Szent-Gyorgyi and J. Blum) rotor produced by DuPont/Sorvall Instruments (Fig. 10-1). The rotor itself is a conventional fixed-angle rotor (model SS-34) and can be used with either two, four, or eight specially fabricated centrifuge tubes. The maximum angular velocity of the assembled rotor system is 20,000 rpm ($RCF_{max} = 48,250g$), and the maximum flow rate is 600 ml/min. The particle suspension to be processed is fed (using either gravity or a pump) into a rather novel *inlet–outlet* assembly mounted on the top of the rotor (Fig. 10-2). The inlet port of this assembly leads into a central, vertical

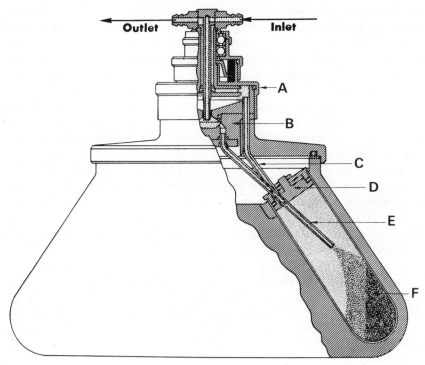

Fig. 10-1 The KSB tube-format CFWP system: *A*, inlet–outlet assembly; *B*, distributor (to two, four, or eight channels); *C*, exit line from tube cap; *D*, tube cap; *E*, entry line through tube cap; *F*, sedimenting particles accumulating in tube. (Courtesy of E. I. DuPont and Company.)

feed line that descends into an axial opening in the *distributor*. The outlet port, which conveys the supernatant from the rotor, communicates with a second and concentric vertical channel in the *lift device;* this channel branches in the circular base of the lift to form two radial channels that open near the wall of the lower section of the inlet–outlet assembly. Two bearings in an upper section of the inlet–outlet assembly permit the central feed line and the concentric lift device to remain stationary while the remaining parts of the assembly and the distributor to which the assembly is attached rotate with the spinning rotor.

During operation of the system, the particle suspension enters the feed line and is deposited in the axial opening of the spinning distributor. Centrifugal force sweeps the suspension into and through steel tubing lines that enter each centrifuge tube aperture in the tube caps (see Fig. 10-1). If the rotor speed and the rate of flow through the rotor are appropriate, particles in the suspension sediment toward and down the sloping outer surface of the centrifuge tubes under the combined influences of convection and centrifugal force and are packed against the bottom of each tube. At the same time, the clarified supernatant is displaced centripetally toward and into the exit lines of the tube caps. The supernatant ascends through the distributor by displacement and enters the lower chamber of the inlet–outlet assembly, where it encounters and enters the two openings at the margins of the stationary lift. The instantaneous change in velocity that occurs as the rotating liquid enters the static lift channels is translated into rapid vertical flow up and out of the rotor through the outlet port.

Particle suspensions may be processed until the sediment accumulating in each tube reaches 35 ml (a total of 280 ml if all eight tube positions in the rotor are used), at which time the rotor must be decelerated and the tubes emptied and cleaned. For a particle suspension containing 10 gm of sediment per liter processed at the maximum flow rate (600 ml/min), this implies that particles in up to 28 l of material can be harvested in a single 47-min operation.

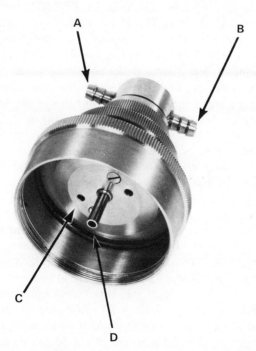

Fig. 10-2 KSB (and TZ-28GK) inlet–outlet assembly: *A*, inlet port; *B*, outlet port; *C*, lift device; *D*, feed line. (Courtesy of E. I. DuPont and Company.)

Whether the processing of a particle suspension using the tube format type of continuous-flow rotor will be acceptable depends principally on the total volume of sediment that one anticipates collecting. Since the KSB system is limited to a maximum sediment volume of 280 ml, any quantity greater than this would require either successive runs or the use of an alternative, higher-capacity format such as a zonal rotor (see below). However, in many applications the total amount of sediment is far less than 280 ml, and it may be more practical to use fewer tube positions in the KSB rotor. Indeed, sediments amounting to only a few grams would be very difficult to recover from the compartments of a zonal rotor or other continuous-flow systems of comparable, large sediment capacity. Small-volume sediments are readily removed from the individual tubes of the KSB system.

Continuous Flow with Pelleting Using Zonal Rotors   Continuous flow with pelleting can also be carried out using zonal rotors adapted for continuous-flow operation (Barringer et al., 1966; Sheeler and Wells, 1972). Among the more popular instruments are the JCF-Z and CF-32Ti rotors of Beckman Instruments (Spinco Division) and the TZ-28GK rotor of DuPont/ Sorvall Instruments (Fig. 10-3). The JCF-Z and CF-32Ti operate in a similar manner, but the JCF-Z uses interchangeable cores: a small-pellet core for trapping up to 190 ml of sediment, a standard core for collecting up to 400 ml of sediment, and a large-pellet core with which up to 800 ml of sediment may be collected. Figure 10-4 shows the relationship for the two Beckman rotors between the sedimentation rates of the particles in the suspension being processed and the maximum allowable flow rates into and out of the rotors. Although the CF-32Ti rotor can be used to collect large-volume sediments, its more common application is for continuous-flow banding of small particles in density gradients, and the rotor is described more fully in that regard later in the chapter.

In the JCF-Z and CF-32Ti a face seal similar to that used for density gradient centrifugation in dynamically unloaded zonal rotors (see Chapter 6) is used to separate the streams of liquid flowing into and out of the rotors. Indeed, the overall size and shape, as well as many other features of the JCF-Z and CF-32Ti rotors, are similar to Beckman Instruments' B-XV zonal rotor. The most notable departure in the configuration of the JCF-Z and CF-32Ti rotors is in the core–septa component. The core diameter is much greater than in a zonal rotor used for density gradient separations and leaves only a small annular space near the bowl wall. After passing through the inlet lines of the rotating seal, the particle suspension is carried into the annular space through four radial channels at the bottom of the core. The supernatant exits the annular space through

Fig. 10-3 The CF-32Ti (left) and TZ-28GK (right) CFWP zonal rotors. (CF-32Ti photograph courtesy of Beckman Instruments, Inc.; TZ-28GK photograph courtesy of E. I. DuPont and Company.)

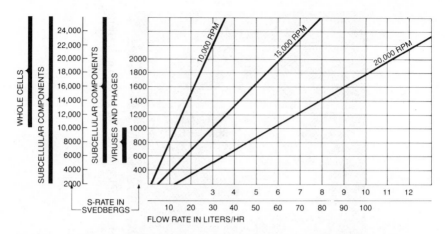

Fig. 10-4 Relationship between particle sedimentation coefficients and allowable suspension flow rates through the CF-32Ti and JCF-Z CFWP zonal rotors at 10,000 rpm, 15,000 rpm, and 20,000 rpm. (Courtesy of Beckman Instruments, Inc.)

another four channels at the top of the core and exits the rotor through the rotating seal. During ascent of the particle suspension through the narrow annular space of the rotor bowl, particles are driven by centrifugal force against the bowl wall. Four narrow, septalike baffles at the edges of the core prevent horizontal swirling of the suspension and the accumulating sediment. Once the requisite amount of sample has been processed, the rotor is decelerated and opened and the sediment scraped from the bowl wall.

The TZ-28GK CFWP rotor is a modified version of the TZ-28 reorienting gradient zonal rotor (see Chapter 6). Substitution of a continuous-flow distributor and inlet–outlet assembly for the density gradient distributor converts the TZ-28 to the TZ-28GK. The inlet–outlet assembly is the same as that used in the DuPont/Sorvall KSB rotor (see above). The continuous-flow distributor serves to mate the feed line of the inlet–outlet assembly (Fig. 10-2) to the six septa lines of the core–septa piece and the exit line and lift device to the six core lines. The particle suspension is carried by centrifugal force through the distributor and septa lines to the vertex of the annular V-shaped groove in the bowl floor. Particles in the suspension sediment up the sloping outer surface of the V groove under the influence of convection and centrifugal force and are packed against the bowl wall. At the same time, the supernatant is displaced centripetally toward and through the exit channels at the top of the core. Ascending through the distributor, the supernatant encounters the lift device in the inlet–outlet assembly and is ejected from the rotor (Fig. 10-5). Unlike the JCF-Z and CF-32Ti rotors, the TZ-28GK does not employ a rotating face-seal system to prevent the two streams of fluid that enter and leave the rotor from mixing; instead, the centrifugal force that exists within the inlet–outlet assembly during operation keeps the two streams apart. After

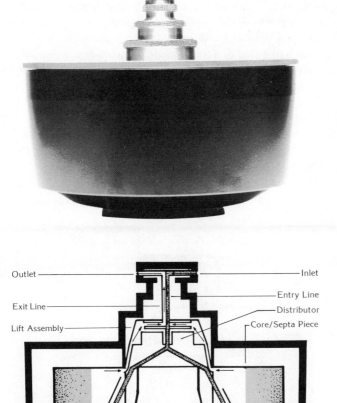

Fig. 10-5 Flow-through scheme of the TZ-28GK CFWP zonal rotor. (Courtesy of E.I. DuPont and Company.)

up to 800 ml of sediment has been collected in the rotor bowl, the TZ-28GK is decelerated, opened, and emptied.

Other CFWP Formats Lourdes Instrument Corporation and MSE Scientific Instruments, Ltd. manufacture CFWP rotors that provide for the collection of the sediment in a large, removable plastic (usually polyethylene or polypropylene) liner. In the Lourdes CFR rotor (Fig. 10-6), the particle suspension is fed through a stationary canister mounted above the spinning rotor and into the rotor's distributor. Centrifugal force sweeps the suspension radially to the edge of the rotor chamber, whose downwardly and outwardly sloping walls act to sediment the particles in a manner much like that of a fixed-angle rotor. While the sediment collects at the centrifugal edge of the chamber, the supernatant is displaced centripetally toward and into an annular set of vertical channels in the rotor cover. These channels lead back up to the canister. As the supernatant emerges from the openings in the rotor cover, centrifugal force sprays the liquid into the surrounding canister. The walls and the sloping base of the canister act simply to convey the liquid to the tube leading from the centrifuge. The Lourdes CFR rotor does not require either a rotating face

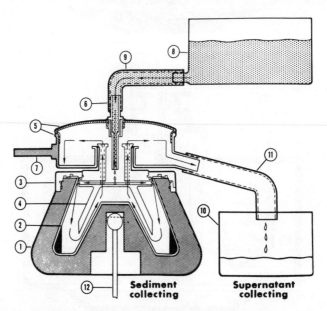

Fig. 10-6 The Lourdes CFWP system: 1, rotor body; 2, liner; 3, cover; 4, discharge cap; 5, canister; 6, inlet tube; 7, support rod; 8, particle suspension; 9, feed line; 10, collection reservoir; 11, discharge line; 12, drive shaft.

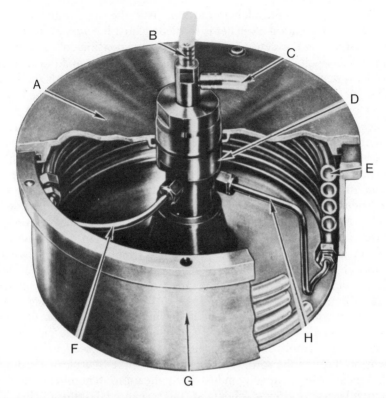

Fig. 10-7 The IEC Helixtractor: *A*, rotor cover; *B*, inlet; *C*, outlet; *D*, distributor assembly; *E*, tubing; *F*, line into helix; *G*, rotor body; *H*, line out of helix.

seal or a lift device; three different models of the rotor accept up to 100 ml, 650 ml, or 1800 ml of sediment. At the conclusion of centrifugation, the rotor is decelerated, the cover removed, and the plastic liner containing the sediment lifted out of the rotor.

An especially novel application of the CFWP approach is that taken in the Helixtractor rotor (International Equipment Co.). In the Helixtractor (Fig. 10-7), the particle suspension is caused to flow in a spiral fashion through a helical coil of plastic tubing stacked against the wall of the hollow, basket-type rotor. One end of the tubing is connected to the inlet channel of the inlet–outlet assembly and the other end to the outlet channel. Since the tubing is spinning with the rotor, particles in the suspension are subjected to centrifugal force that is directed *across the diameter* of the tubing as the suspension simultaneously advances along the tubing's length. All material passing through the tubing is subjected to the same RCF since the entire length of tubing is at the same radial distance from the axis of rotation. A pellet is formed against the wall of the tubing with the largest and/or densest particles of the suspension collected in the length of tubing closest to the incoming flow of material and the smallest and/or least dense particles deposited further on. In other words, particles in the suspension are *differentially* deposited along the length of the tubing in order of decreasing sedimentation coefficient.

Once the suspension has been processed, the rotor is decelerated and the tubing removed. Either the entire sediment may be removed by expelling it from the tubing using a roller extractor; or if a specific phase or portion of the sediment is desired (the sediment is visible through the transparent wall of the plastic tubing), it may be isolated by cutting out that particular section of tubing from the overall length. The inlet–outlet assembly of the Helixtractor rotor operates on principles that are similar to those in effect in the KSB system described earlier in the chapter.

For many years, the Sharples Centrifuge Company (now a division of Pennwalt Corporation) has produced a variety of motor-driven and turbine-driven continuous-flow centrifuges in which the particle suspension is fed into the narrow tubular rotor through an axial opening in the rotor's floor while the supernatant is discharged through one or more ports at top of the rotor (Fig. 10-8). Discharge takes place entirely by overflow of the rotor, since external pressure is not used to drive the suspension through the rotor. Although many of the Sharples centrifuges are designed for high-capacity industrial applications, a number of models are in regular use in clinical, chemical, and research laboratories (Fig. 10-9). Depending on the nature of the particle suspension being processed and the choice of rotor (a selection of application-specific rotors is available), the sediment can be trapped on the wall of the rotor while the supernatant is separately discharged as dense and light phases. The dis-

Table 10-1 CFWP Zonal Rotors

Rotor	Sediment Capacity (ml)	Maximum rpm	Maximum RCF	Maximum RCF at Entry	Maximum Flow Rate (l/hr)
JCF-Z (Beckman	400[a]	20,000	40,000g	32,000g	45
CF-32Ti (Beckman)	305	32,000	102,000g	86,000g	9
TZ-28GK (DuPont/Sorvall)	800	20,000	43,000g	30,000g	36

[a] Standard core; small-pellet (190-ml) and large-pellet (800-ml) cores are also available.

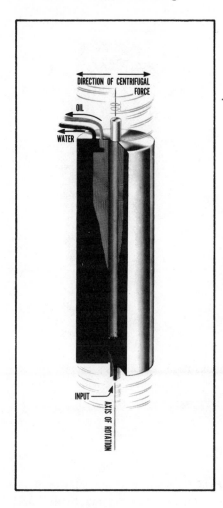

Fig. 10-8 Continuous-flow system used in the Sharples tubular rotors. (Photograph courtesy of Sharples-Stokes Division, Pennwalt Corporation.)

charged phases enter different covers mounted above one another at the top of the rotor and are separately carried away. The principle applied is reminiscent of that used in the early centrifugal cream separators pioneered by de Laval in the late nineteenth century. In certain applications, the supernatant may be vented through one port while the sediment is continuously discharged as a sludge through the other port, thus allowing the rotor to process much larger volumes of suspension before it is decelerated and emptied.

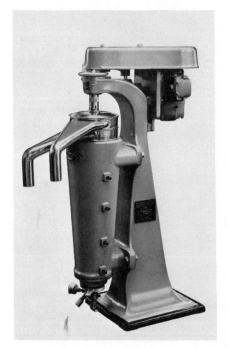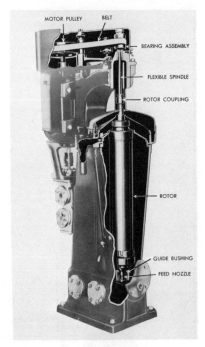

Fig. 10-9 Sharples tubular rotor continuous-flow centrifuges (see text for description). (Photographs courtesy of Sharples-Stokes Division, Pennwalt Corporation.)

CONTINUOUS FLOW WITH BANDING

The CFWB approach to continuous-flow harvesting of particles evolved in response to the observations that certain viruses are inactivated and that many subcellular particles form intractable aggregates when they are pelleted in a CFWP rotor. In a CFWB rotor, the suspension is caused to flow over the surface of a *density gradient* imprisoned in the spinning rotor. Particles are sedimented out of the stream and are banded isopycnically in the gradient while the particle-free supernatant is conveyed from the rotor. In this manner the particles accumulated in the rotor are maintained in suspension and retain their activity. Moreover, with this technique, buoyant density as well as sedimentation coefficient can be used as a parameter for separating and harvesting a specific population of particles. Continuous-flow-with-banding rotors were an outgrowth of the Zonal Centrifuge Development Program (later the Molecular Anatomy Program or the MAN program) of the Oak Ridge National Laboratory directed by N. G. Anderson (see Chapter 6), and essentially all modern commercial versions of CFWB rotors are based on the designs and the operating principles of the Anderson forerunners (see Anderson, 1968).

Origins of Current CFWB Rotors   Beginning with a modified version of the now-classic B-IV zonal rotor, Anderson tested a succession of designs for continuous-flow centrifugation. The first of these rotors, the B-V was designed for pelleting the suspended particles against the rotor wall (Barringer et al., 1966), but the B-VI, B-VII, B-VIII, and B-IX versions were altered so that the particles could be banded in a density gradient (Anderson et al., 1966; Anderson, 1968). In the same general way that the configuration of the B-IV zonal rotor was subsequently altered to produce the lowered profile and the highly successful B-XV, a similar change in configuration lead from the B-VIII to the B-XX CFWB rotor. The B-XX rotor is produced commercially at this time by MSE (designated model B-XX) and by Beckman Instruments (designated CF-32Ti). Both rotors may be used for pelleting, but particle banding in density gradients is the primary function.

Operation of CFWB Rotors   The Beckman Cf-32Ti rotor appears in Fig. 10-3 and its various parts are shown diagrammatically in Fig. 10-10. Operation of this rotor is representative of CFWB rotors in general. The large diameter of the rotor's core leaves only a small annular space in the bowl (i.e., 430 ml) for the density gradient and particle suspension (usually 305 ml of gradient and 125 ml of suspension). Septa attached to the edges of the core divide the annular space into sector-shaped compartments and prevent horizontal circulation of the rotor's contents. The core contains two sets of radial channels for the continuous flow of sample through the rotor, and four slots milled into the top surface form another set of passages used to initially fill and later empty the rotor. The core is tapered; the bottom has a slightly smaller radius than the top. The radial channels at the bottom of the core converge on a single axial line, whereas the radial channels in the top of the core and the four milled slots converge to form an annular line. The axial and annular lines communicate with the exterior of the rotor through a rotating face seal similar to that used in a dynamically unloaded zonal rotor such as the B-XV.

   With the empty rotor spinning at low speed, the density gradient is pumped light end first into the annular line and flows to the wall of the

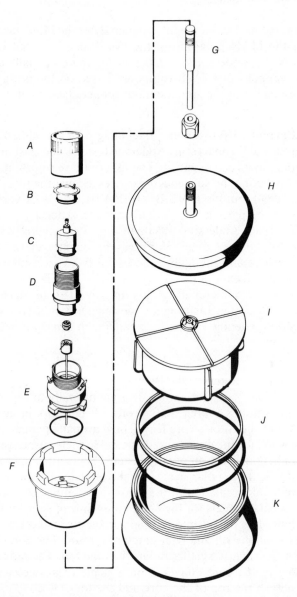

Fig. 10-10 Components of the CF-32Ti CFWB zonal rotor (see also Fig. 10-3): *A*, collar; *B*, manifold; *C*, stationary seal; *D*, water jacket; *E*, rotating seal and bearing housing; *F*, bowl adapter; *G*, stem; *H*, rotor lid; *I*, core–septa assembly; *J*, nonextrusion ring and gasket; *K*, rotor bowl. (Courtesy of Beckman Instruments, Inc.)

rotor through the milled slots; at the same time, air is displaced from the rotor through the axial line. With the gradient loaded and the rotor now spinning at the processing speed (up to 32,000 rpm), the suspension is pumped through the axial line to the bottom of the core and radially onto the light end of the gradient. The flow rate is adjusted so that as the suspension ascends through the rotor over the surface of the gradient the particles of interest have the opportunity to enter and become trapped in the gradient. The supernatant flows out through the radial channels at the top of the core and exits the rotor through the annular line. To collect the banded particles at the end of the run, an air block is first introduced into the annular line to prevent liquid from traveling through the radial channels at the top of the core, and the dense cushion is pumped to the rotor wall through the milled slots at the top of the core. The gradient and the entrained particles are thus gradually displaced toward and into the radial channels at the bottom of the core and up and out of the rotor through the axial line.

Unlike the rotating face seal of a zonal rotor such as the B-XIV, B-XV and B-XXIX models, the rotating seal of the CF-32Ti remains in position throughout the run. Consequently, a special lubricating and cooling apparatus is associated with the seal (see Fig. 10-10). The CF-32Ti (and the MSE B-XX) rotor is designed to operate in an ultracentrifuge; a somewhat simpler form of the CF-32Ti, the Beckman JCF-Z rotor described earlier in the chapter in connection with CFWP, can be operated in superspeed centrifuges.

Yet another series of CFWB rotors based on Anderson designs are the K-type rotors used for commercial or other large-scale harvesting of virus particles (Reimer et al., 1967; Perardi et al., 1969; Cline and Dagg, 1973). These rotors are operated in air-turbine centrifuges. In K-type CFWB rotors, the single coaxial face seal is replaced by separate entry and exit lines axially positioned at the top and the bottom of the rotor; these lines communicate with the annular banding chamber through radial channels in the tapered rotor core. Although dynamic loading and unloading are possible, the density gradient is usually introduced with the rotor at rest and reoriented from the vertical to the radial position during acceleration. Similarly, the gradient and the entrained particle bands are collected from the rotor after deceleration and reorientation.

The DuPont/Sorvall SZ-14GK and TZ-28GK zonal rotors are designed principally for CFWP applications. However, Nixon et al. (1973) described a procedure for carrying out CFWB operations in these rotors, where the particle bands are collected after deceleration of the rotor and reorientation of the density gradient.

CENTRIFUGAL ELUTRIATION

Centrifugal elutriation (or *counterstreaming centrifugation*) is a special type of continuous-flow centrifugation that can be used to separate populations of particles that have different sedimentation characteristics. Although the technique was pioneered in the late 1940s and early 1950s by P. E. Lindahl (Lindahl, 1948; Lindahl, 1956), it has since been brought to its present state of the art primarily through the efforts of C. R. McEwen and E. T. Juhos (McEwen et al., 1968; Glick et al., 1971).

The principles of elutriation are quite simple and may best be explained by example. If water enters the narrow end of a vertically positioned conical chamber (such as a funnel) at a *constant* volume per minute, then the rate at which the liquid flows upward through the chamber *diminishes* with distance from the chamber's entry port. This is because the cross-sectional area of the chamber continuously increases. We will suppose that as the liquid overflows the top of the chamber, it is diverted to a fraction collector. Now, if the liquid in the chamber contains a mixture of suspended particles varying in size and density, each particle experiences two *opposing* forces: (1) the force due to gravity causing the particles to sediment toward the chamber's entry port and (2) the force due to the liquid flow in the opposite direction. If at a given level in the chamber a particle's sedimentation rate is greater than the rate of liquid flow at that level, the particle will experience a net movement toward the entry port of the chamber; whereas if the flow rate exceeds the particle's sedimentation rate, the particle will be carried in the direction of the flow. Since the flow rate varies through the height of the chamber, then for each particle there will be a level at which the two forces exactly cancel one another, and once reaching this level (by either moving upward or downward) the particle will undergo no further migration. Different populations of particles would form layers at different levels of the chamber. (All this necessitates that the volume of liquid entering the chamber per minute not be so great that it carries the particles out of the chamber or so small that the particles pile up at the entry port; such conditions are readily avoided.)

If after the populations of particles have formed separate layers, the volume of liquid entering the chamber per minute is *increased,* all the layers will be displaced vertically. A succession of increases in input volume would serially displace the layers through the top of the chamber, where they may be separately collected.

In this example, the force on the particles that opposes the flow of liquid is gravity and is constant. In centrifugal elutriation, the particle separation chamber is in a spinning rotor. This not only increases the tendency of the particles to sediment (thereby reducing the time necessary for particles to reach equilibrium positions), but since the RCF varies with radial position, another variable is introduced to the process. It is to be noted that the magnitude of the centrifugal force and the flow rate vary in *opposite* directions during centrifugal elutriation.

THE ELUTRIATOR ROTOR

Lindahl's original centrifugal elutriator was a rather complex apparatus and was never produced commercially. However, using Lindahl's concept, Beckman Instruments began development of a simpler device in the 1960s. Early versions (rotors CR-1 and CR-2) lead to the JE-6 rotor that is now available commercially and that can be used in superspeed centrifuges at speeds of up to 6000 rpm. The assembled JE-6 rotor is shown in Fig. 10-11 and the separation chamber and bypass chamber (which acts as a counterpoise) in Fig. 10-12.

The solid plastic rotor body contains two openings for insertion of the separation chamber and bypass chamber and the conduits leading to and from the core. The core is inserted through an axial opening in the rotor body and contains entry and exit lines that communicate with the rotor body's conduits and also with the exterior of the centrifuge through a

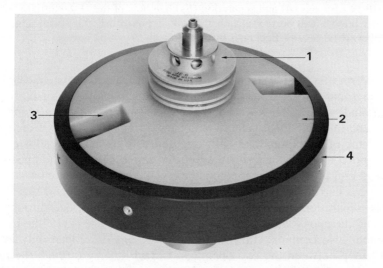

Fig. 10-11 The JE-6 elutriator rotor: 1, rotating seal assembly and bearing; 2, rotor body; 3, recess for separation chamber; 4, pressure ring. (Courtesy of Beckman Instruments, Inc.)

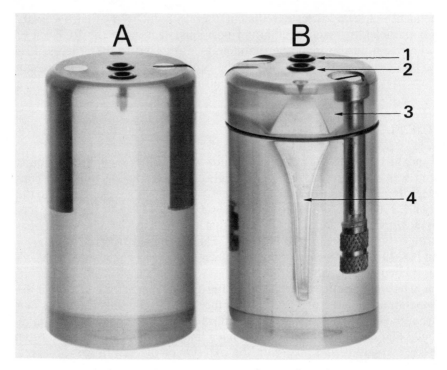

Fig. 10-12 Bypass (*A*) and separation (*B*) chambers of the JE-6 elutriator rotor: 1, entry port; 2, exit port; 3, chamber cap; 4, exponentially flared cell-separation chamber. (Courtesy of Beckman Instruments, Inc.)

coaxial rotating-seal assembly much like that used in a dynamically unloaded zonal rotor. A metal pressure ring encircles the rotor body and aligns and seals the separation and bypass chambers against the conduits. The separation chamber is fabricated from a transparent epoxy and takes the shape of an exponential horn oriented with its narrow end (i.e., the *entry* port) at the rotor's edge. An epoxy cap attached to the open centripetal end of the separation chamber reconverges the chamber's walls to form a narrow *exit* port. Liquid makes its way to the centrifugal edge of the separation chamber through a narrow channel drilled through the cap and the epoxy wall of the separation chamber. The cap of the bypass chamber also contains entry and exit ports, but these lead only into a short loop between the two openings. Liquid flowing through the rotor first passes through the separation chamber and then the bypass chamber. The bypass chamber and the flow of liquid through it during operation of the rotor act to counterbalance the separation chamber, whose capacity is about 4.5 ml. A stroboscopic lamp is mounted in the centrifuge chamber below the rotor and can be synchronized with the rotor's speed. Thus, with the standard door of the centrifuge replaced with a modified door containing a viewing port, the progress of elutriation in the JE-6's transparent separation chamber can be followed.

OPERATION OF THE ELUTRIATOR ROTOR

Prior to acceleration, the rotor and all connecting lines are filled with water or buffer solution. The rotor is then accelerated and the mixture of particles to be separated pumped through the rotating seal and into the separation chamber (Fig. 10-13). With the sample positioned in the separation chamber, water or buffer is slowly pumped through the separation chamber. By carefully adjusting the rotor speed and the liquid flow rate, the population of particles with the lowest sedimentation rate can be flushed from the chamber. This can be monitored visually through the centrifuge door. Usually, 100 ml of liquid is required to flush one population of particles from the 4.5-ml separation chamber. Either increasing the flow rate by a factor of 1.5 or decreasing the rotor speed by 25% is usually adequate to flush successive fractions of particles from the elutriator. The fractions collected from the rotor represent particles of greater and greater sedimentation coefficient. The separation of two hypothetical populations of particles by centrifugal elutriation is depicted in Fig. 10-13. Once all the particles have been displaced, the rotor is halted.

To date, the elutriator rotor has been used primarily for whole-cell separations (see reviews by Sanderson and Bird, 1977 and Grabske, 1978) and has been used successfully with blood cells (McEwen et al., 1968), mast cells from peritoneal fluid (Glick et al., 1971), brain cells (Flangas, 1974), testis tissue cells (Grabske et al., 1975), liver tissue (Knook and Sleyster, 1976), and yeasts (Gordon and Elliot, 1977). By using the elutriator to selectively isolate cells of similar size (and therefore also similar age) from cell cultures and by using the isolate as an inoculum, synchronous cell cultures can be obtained (Gordon and Elliott, 1977; Meistrich et al., 1977). Since a density gradient is not used in the centrifugal elutriator to stabilize the zones of particles that form in the separation chamber, it is important that the particle concentration not be so high as to provoke streaming and inversion. For whole-cell separations, samples containing 10^7 to 10^8 cells can be used without overloading the separation chamber.

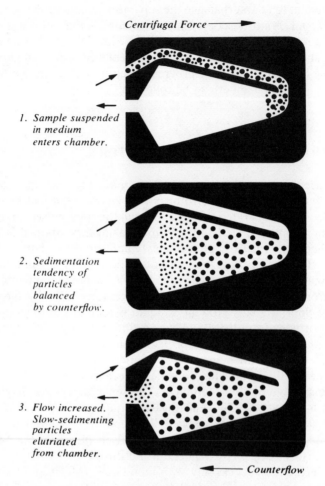

Fig. 10-13 Separation of two hypothetical populations of particles in the JE-6 elutriator rotor. (Courtesy of Beckman Instruments, Inc.)

CONTINUOUS-FLOW PLANETARY MOTION CENTRIFUGE

In CFWB and Anderson-type dynamically unloaded zonal rotors, the flow of liquid between the stationary tubing lines mounted external to the spinning rotor and the channels within the rotor itself occurs across the faces of the rotating seal. In 1966, Ito et al. described the application of a unique and novel principle that makes it possible to eliminate face seals while still maintaining continuous (i.e., uninterrupted) tubing connections between a spinning rotor and stationary feed/collection lines. Ito's system is referred to as a *planetary motion, flow-through centrifuge* (Ito and Bowman, 1971 and 1974; Ito et al., 1975).

In the planetary motion, continuous-flow centrifuge, the inlet and outlet channels of the rotor bowl converge on ports positioned close to the axis of rotation and on the undersurface of the rotor. Flexible tubing lines (two or more), gathered together to form a bundle, are connected to these ports and loop up and around the rotor bowl through a hollow support rod and exit the centrifuge above the axis of the rotor. If the rotor bowl rotates around its axis at an angular velocity of 2ω, while the hollow rod containing the bundle of flexible tubing rotates around the same axis at ω, then the tubing lines remain free and do not twist around one another. To achieve this, the tubing bundle must rotate about its own axis at $-\omega$ (i.e., opposite to the direction of rotation of the rotor bowl). The synchronous rotational speeds of the rotor bowl and the tubing lines are maintained by sets of gears and pulleys that link the rotor bowl with the hollow tubing support rod. It is the rotation of the tubing support rod (and the enclosed bundle of tubing lines) around its own axis as well as around the rotor bowl that constitutes the planetary motion. The centrifuge has been used successfully to separate whole blood into cellular and plasma fractions on a continuous-flow basis (Ito et al., 1975).

In several interesting variations of the centrifuge, a sample-filled helical chamber replaces the tubing support rod. The planetary motion experienced by particles suspended in the chamber serves to sort them into separate populations distributed along the length of the helix; the separations may be based on differences in particle weights, densities, or other properties. After the particles have been separated, they are serially displaced from the helical chamber (Ito and Bowman, 1971, 1978; Ito et al., 1979).

REFERENCES AND RELATED READING

Books

Anderson, N. G. (Ed.) *The Development of Zonal Centrifuges and Ancillary Systems for Tissue Fractionation and Analysis,* National Cancer Institute Monograph Series, No. 21, Bethesda, Maryland, 1966.

Birnie, G. D., and Rickwood, D. (Eds.) *Centrifugal Separations in Molecular and Cell Biology.* Butterworths, London, 1978.

Articles and Reviews

Anderson, N. G. (1967) Preparative zonal centrifugation. In *Methods of Biochemical Analysis,* D. Glick, Ed. Wiley-Interscience, New York.

Anderson, N. G. (1968) Preparative particle separation in density gradients. *Quart. Rev. Biophys.,* **1,** 3.

Anderson, N. G., Barringer, H. P., Amburgey, J. W., Cline, G. B., Nunley, C. E., and Berman, A. S. (1966) Continuous-flow centrifugation combined with isopycnic banding: Rotors B-VIII and B-IX. *National Cancer Inst. Monograph 21,* p. 199.

Barringer, H. P., Anderson, N. G., and Nunley, C. E. (1966), Design of the B-V continuous-flow centrifuge system. *National Cancer Inst. Monograph 21,* p. 191.

Cline, G. B., and Dagg, M. K. (1973) Particle separations in the J- and RK-types of flo-band zonal rotors. In *Methodological Developments in Biochemistry,* Vol. 3, *Advances with Zonal Rotors,* (E. Reid, Ed.) Longmans Group Publishers, London.

Flangas, A. L. (1974) Bulk separations of rat brain cells by centrifugal elutriation. *Prep. Biochem.,* **4,** 165.

Glick, D., von Redlich, D., Juhos, E. T., and McEwen, C. R. (1971) Separation of mast cells by centrifugal elutriation. *Exp. Cell Res.,* **65,** 23.

Gordon, C. N., and Elliott, S. G. (1977) Fractionation of *Saccharomyces cerevisiae* cell populations by centrifugal elutriation. *J. Bacteriol.,* **129,** 97.

Grabske, R. J. (1978) Separating cell populations by elutriation. *Fractions,* No. 1, 1.

Grabske, R. J., Lake, S., Gledhill, B. L., and Meistrich, M. L. (1975) Centrifugal elutriation: Separation of spermatogenic cells on the basis of sedimentation velocity. *J. Cell. Physiol.,* **86,** 177.

Grant, W. D., and Morrison, M. (1979) Resolution of cells by centrifugal elutriation. *Anal. Biochem.,* **98,** 112.

Ito, Y., Weinstein, M., Aoki, I., Harada, R., Kimura, E., and Nunogaki, K. (1966) The coil planet centrifuge. *Nature,* **212,** 985.

Ito, Y., and Bowman, R. L. (1971) Countercurrent chromatography with flow-through coil planet centrifuge. *Science,* **173,** 420.

Ito, Y., and Bowman, R. L. (1974) Particle separation with the flow-through coil planet centrifuge. *Anal. Biochem.,* **61,** 288.

Ito, Y., Suadeau, J., and Bowman, R. L. (1975) New Flow-through centrifuge without rotating seals applied to plasmapheresis. *Science,* **189,** 999.

Ito, Y. and Bowman, R. L. (1978) Countercurrent chromatography with the flow-through centrifuge without rotating seals. *Anal. Biochem.,* **85,** 614.

Ito, Y., Carmeci, P., and Sutherland, I. A. (1979) Nonsynchronous flow-through coil planet centrifuge applied to cell separation with physiological solution. *Anal. Biochem.,* **94,** 249.

Knook, D. L., and Sleyster, E. C. (1976) Separation of Kupffer and endothelial cells of the rat liver by centrifugal elutriation. *Exp. Cell Res.,* **99,** 444.

Lindahl, P. E. (1948) Principle of a counterstreaming centrifuge for the separation of particles of different sizes. *Nature,* **161,** 648.

Lindahl, P. E. (1956) On counterstreaming centrifugation in the separation of cells and cell fragments. *Biochim. Biophys. Acta,* **21,** 411.

McEwen, C. R., Stallard, R. W., and Juhos, E. T. (1968) Separation of biological particles by centrifugation elutriation. *Anal. Biochem.,* **23,** 369.

Meistrich, M. L., Meyn, R. E., and Barlogie, B. (1977) Synchronization of mouse L-P59 cells by centrifugal elutriation separation. *Exp. Cell Res.,* **105,** 169.

Nixon, J. C., McCarty, K. S., and McCarty, K. S. (1973) The use of a reorienting density gradient rotor with continuous sample flow for the isolation of calf thymus nuclei. *Anal. Biochem.,* **55,** 132.

Perardi, T. E., Leffler, A. A., and Anderson, N. G. (1969) K-series centrifuges. II. Performance of the K-II rotor. *Anal. Biochem., 32,* 495.

Reimer, C. B., Baker, R. S., van Frank, R. M., Newlin, T. E., Cline, G. B., and Anderson, N. G. (1967) Purification of large quantities of influenza virus by density gradient centrifugation. *J. Virol.,* **1,** 1207.

Sanderson, R. J. and Bird, K. E., Cell separations by counterflow centrifugation. In *Methods in Cell Biology,* Vol. XV, D. M. Prescott, Ed. Academic, New York, 1977.

Sheeler, P., and Wells, J. R. (1972) Continuous-flow centrifugation. *Amer. Lab.* (January), 62.

ELEVEN

Separation of Particles Using Unit Gravity Devices

The separation of particles on the basis of differences in their sedimentation coefficients or densities may not require the use of centrifugation if the particles are sufficiently large. Among such particles are *whole cells* which sediment fairly rapidly even at $1 \times g$—that is, when they are subjected only to the earth's gravitational force. The use of "unit gravity" techniques to separate different cell types has become increasingly popular in the last several years, and there has been much success with bone marrow cells (Zeiller and Hansen, 1978, 1979), spermatogenic cells (Meistrich, 1977), ascites fluid cells (Bont and Hilgers, 1977; Sheeler and Doolittle, 1980), hemopoietic cells (Mel and Mohandas, 1979), liver cells (Tulp and Bont, 1975), and kidney cells (Tait et al., 1974).

It is not really surprising that so simple an approach as using the earth's gravity to affect a separation of different particles in a density gradient is so recent a development and area of intense interest. During the first half of this century, attention was focused for the most part on the isolation and the fractionation of small particles (e.g., macromolecules) and microscopic organelles from tissue homogenates. The earth's gravity is simply far too small a force to effectively sediment these particles, and centrifugation becomes essential. Until recently, the cells present in tissues used to prepare homogenates were considered to be sufficiently similar that one could justify subcellular analyses by presuming that the particles isolated were released from a uniform cell population. With the development of methods to dissociate tissues (e.g., by organ perfusion prior to extirpation) into their separate cell components, it became possible to examine different metabolic activities or properties as they are distributed among the multiplicity of discrete cell types that comprise most tissues.

The sedimentation coefficients of most cells are very high (e.g., $10^7 S$ to $10^8 S$) and the use of centrifugation to separate the different types of cell present in a heterogeneous mixture poses special problems. Even at modest angular velocities, the sedimentation rate of cells is so great that they may move to the bottom of the centrifuge tube or to the wall of a zonal rotor in just a few seconds or a few minutes. The use of centrifugal force for such separations has relied for the most part on the use of rotors with especially short minimum and maximum radii or the attenuation of the cells' sedimentation rates by using high-density (or viscosity) gradients. These options introduce additional restrictions and problems (see below), so that the use of the unit gravity approach has become the most broadly accepted alternative.

THE "STA-FLO" SYSTEM OF H. C. MEL

The first successful use of unit gravity to separate different types of cells was that achieved in the "stable-flow free boundary" (Sta-Flo) system of H. C. Mel (Mel, 1963, 1964). Mel's apparatus consisted of a long, narrow rectangular chamber through which as many as 12 parallel layers of sucrose solutions of different densities were passed. The gradient steps were introduced through ports at one end of the chamber and collected at the other end, and the "residence time" was about 30 minutes (Fig. 11-1). The density steps, together with the slow and smooth flow rates, resulted in little or no mixing between adjacent layers. The cell sample was continuously introduced into the top sucrose layer at one end of the chamber. Depending on (1) the sedimentation coefficients of the cells in each density layer and (2) the lateral flow rates of the gradient steps, each cell type followed a specific arc-shaped, descending path through the chamber (Fig. 11-1). Cells present in different density layers at the ends of their residence times were collected as separate fractions. In its original form, Mel used the technique to successfully separate nucleated and non-nucleated bone marrow cells (Mel, 1963). A recent review by Mel and Mohandas (1979) provides additional descriptions of the Sta-Flo approach, new innovations in the design of apparatus, and further applications to cell sorting.

"STA-PUT" SYSTEMS

Whereas the Sta-Flo approach pioneered by Mel has not achieved widespread use, a gross simplification of the basic rationale called "sta-put"

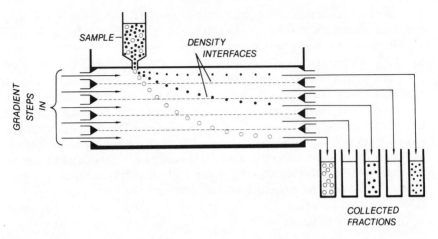

Fig. 11-1 H. C. Mel's Sta-Flo procedure for separating particles at unit gravity. See text for description.

has become quite popular as a tool for separating cells and comparably sized particles at unit gravity. In a sta-put system the sedimentation chamber (rectangular or cylindrical) is filled with a shallow and stationary (i.e., nonflowing) continuous density gradient, the cell suspension carefully layered over the surface, and the cells permitted to fall through the gradient to form separate layers that are then collected for further study. There are a number of different sta-put systems in use, and several of these are described later in the chapter, but at this point it is appropriate to consider the special factors that govern the behavior of cells falling at $1 \times g$ through a density gradient and to compare these with the behavior of cells sedimenting through gradients in rotors under the influence of centrifugal force.

PARTICLE SEDIMENTATION BEHAVIOR IN UNIT GRAVITY SYSTEMS

In contrast to centrifugation, the factors that govern the behavior of particles sedimenting at unit gravity are considerably simpler. In a spinning rotor the force experienced by a particle increases in proportion to the particle's distance from the axis of rotation, whereas in unit gravity devices the force on the particle remains constant and is equal to

$$F = mg \qquad (11\text{-}1)$$

where g is the earth's gravitational acceleration constant and may be taken to be 980 cm sec^{-2}. Two forces oppose the falling particle: the buoyant force F_B of the surrounding medium and the force of friction F_f, the frictional resistance. Unlike their counterparts in centrifugation, these forces are also more simply defined (see also Chapter 2).

$$F_B = \frac{m}{\rho_P} (\rho_M)g \qquad (11\text{-}2)$$

$$F_f = 6\pi\eta r \frac{dx}{dt} \qquad (11\text{-}3)$$

In these equations the particles are assumed to be spherical with radius r, mass m, and density ρ_P, falling at the rate dx/dt through a liquid of density ρ_M and viscosity η.

If the density and the viscosity of the liquid remain constant (i.e., if there is no density gradient), then the falling particle attains a limiting (and constant) velocity when $F = F_B + F_f$ or when

$$mg = \frac{m}{\rho_P} (\rho_M)g + 6\pi\eta r \frac{dx}{dt} \qquad (11\text{-}4)$$

Substituting $\frac{4}{3}\pi r^3 \rho_P$ (i.e., the mass of a particle is equal to the product of its volume and density) for m in equation 11-4 and solving for dx/dt, we obtain

$$\frac{dx}{dt} = \frac{2r^2(\rho_P - \rho_M)}{9\eta} g \qquad (11\text{-}5)$$

Equation 11-5 is equivalent to the sedimentation rate equation except that g replaces $\omega^2 x$. Unlike centrifugation in which dx/dt may continually increase as x increases, dx/dt cannot increase continuously when unit gravity methods are used. Furthermore, since a zone of particles cannot stably sediment through a liquid of uniform density (i.e., a density gradient is mandatory) and thus $(\rho_P - \rho_M)/\eta$ continually diminishes, the velocity at which particles sediment during unit gravity experiments must decrease. Isokinetic conditions can never exist, although they may be closely approximated if the slope of the density gradient is minimized (Fig. 11-2).

For a mixture of different-sized particles that all have the same density, the effectiveness of gradient separations depends on a consideration of how size differences affect sedimentation rates at each level of the density gradient. The particles begin sedimenting together, form separate zones for some period of time, and then close on one another again as the isopycnic position is approached (Fig. 11-3). The separation between any two populations of particles (e.g., curves A and B in Fig. 11-3) reaches

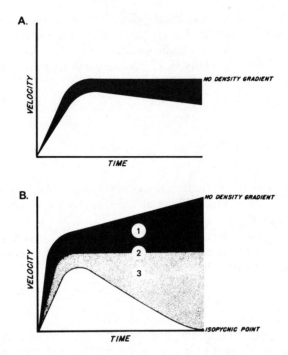

Fig. 11-2 Comparison of the relationship between velocity of sedimentation and time at $1 \times g$ (A) and during centrifugation (B). After an initial acceleration from rest at $1 \times g$, a particle either sediments isokinetically (i.e., if no gradient is present) or its sedimentation rate continuously decreases (dark zone). The shallower the density gradient, the more closely the sedimentation rate approximates isokinetic conditions. In contrast to this, when subjected to centrifugation a particle may undergo three alternative states of velocity change, depending on the gradient's density slope and viscosity; these are (1) continuous acceleration, (2) constant velocity (isokinetic sedimentation), and (3) continuous deceleration.

a maximum when their rates of sedimentation become equal, and this is the point in time at which the separation should be terminated.

Mixtures of biological particles are usually characterized by differences in *both* size *and* density. In the case of particles that are amenable to separation by unit gravity procedures—primarily whole cells—size differences are usually far greater than density differences. This is true not only in the case of cultures of cells, but also in naturally occurring, mixed populations (peripheral blood cells, bone marrow aspirations, dispersed or dissociated tissues, etc.). The densities of most cells fall in the range 1.06 to 1.12, whereas cell radii vary from about 3 to 10 μ. Although sedimentation rate equations clearly show that size more pronouncedly influences sedimentation coefficients than does density, the separation of cells by unit gravity methods must take into account the density differences that exist. Although there are many examples of cells at the small–light end of the size–density spectrum and still others at the large–dense end of the spectrum, in many tissues (and even in cell cultures) an inverse relationship exists between cell size and density, with the result that small and large cells may sediment at similar rates in a certain range of gradient densities. An illustration of this is given in Table 11-1 for two hypothetical particles of radius 5 and 12 and density 1.12 and 1.06, respectively. It may be seen that the particles sediment at appreciably different rates only when the gradient density is either very low or close to the isopycnic density of the less dense particle. Since particle separations at unit gravity are normally quite slow, gradients of low limiting densities are to be preferred (and even then it may be necessary for the particle sedimentation to proceed for several hours). Shallow gradients extending over the density limits that are commensurate with particle zone stability (i.e., gradients that will support the sample zone and prevent streaming and band inversion) provide the most effective cell separations. Such gradients are especially sensitive to temperature perturbations and

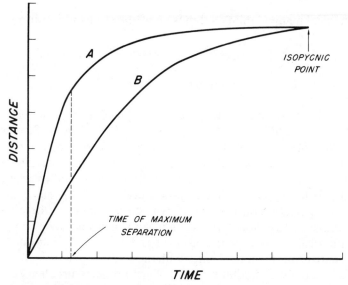

Fig. 11-3   For a mixture of two populations of particles of the same density but different sizes, separation reaches a maximum when their sedimentation rates become equal (compare with Fig. 5-1).

Table 11-1 Comparison of the Sedimentation Rates of Large, Low-Density Particles and Small, High-Density Particles Through Solutions of Various Densities

	dx/dt^a	
ρ_M	Particle A, $r = 5$, $\rho_P = 1.12$	Particle B, $r = 12$, $\rho_P = 1.06$
1.00	3.00	8.64
1.01	2.75	7.20
1.02	2.50	5.76
1.03	2.25	4.32
1.04	2.00	2.88
1.05	1.75	1.44
1.055	1.63	0.72
1.06	1.50	0^b

[a] Expressed in relative units.
[b] Isopycnic for particle B.

to mechanical manipulation, and these must be stringently controlled (see below).

TYPES AND OPERATION OF STA-PUT DEVICES

The first, simple sta-put devices were introduced in the late 1960s by R. G. Miller, R. A. Phillips, E. A. Peterson, and W. H. Evans (Peterson and Evans, 1967; Miller and Phillips, 1969; Brubaker and Evans, 1969). Since then, a large number of variations and modifications have been tested and applied (e.g., Tait et al., 1974; Tulp and Bont, 1975; Zeiller et al., 1976; Bont and Hilgers, 1977; Sheeler and Doolittle, 1980), since most researchers especially construct their own sta-put devices (there are only a few commercial versions). In its simplest form, a sta-put device consists of a cylindrical chamber with a conical end cap attached at one or both ends. A narrow port communicates with the central opening in the cone(s), and this is usually baffled in some manner to reduce turbulence as liquid is either pumped into or withdrawn from the cylinder. The total number of particles that can be safely loaded onto a shallow, low-density gradient that remains at rest (i.e., remains at $1 \times g$) is considerably smaller than during density gradient centrifugation. For this reason and for other practical considerations, sta-put chambers usually have a large capacity (typically one or more liters).

In a typical experiment the sample is placed in the bottom conical end cap and carefully floated upward as the density gradient is slowly introduced (light end first) below. As the sample rises through the end cap, it increases in surface area, spreading the cells over the light end of the gradient as a thin layer. The sedimentation of cells (or other particles) begins even before the chamber is completely filled with gradient. In view of the lengthy period generally required for particle separations at $1 \times g$, this approach does save some time. Sta-put chambers are usually constructed of glass or Lucite so that the initial sample zone and the separating particle zones can be observed and monitored by the operator. Although separations can be carried out at room temperature, it is more common to enclose the separation chamber in a cold-water bath or circulate coolant through an enclosing jacket to preserve cell viability or biological activity.

Once the particles have been separated into discrete layers, the gradient is collected through the port in the lower cone; alternatively, if there is an upper cone, the gradient may be displaced up and out of the chamber by using a dense cushion.

Various modifications of the basic design have been tried in an effort to either improve resolution or increase the numbers of particles that can be separated in a single run. Many of these modifications have focused on the advantages of producing as narrow (as well as uniform) a starting zone as possible and, at the conclusion of the run, collecting the separated zones with a minimum of zone broadening. For example, including polyethylene oxide (or some other light but mildly viscous material) in the sample appears to appreciably increase the number of particles that can form a stable layer on the surface of the gradient without particle aggregation, localized density inversions, and the resulting streaming effect. Placement of a wire screen over the gradient surface prior to application of the sample to evenly disperse the particles has also been tried and found effective. The screen is carefully removed after the particles have descended into the gradient (Tulp and Bont, 1975).

In an elegant sta-put device designed and tested by K. Zeiller and his colleagues (Zeiller et al., 1976), the suspension of particles is introduced onto the gradient through 25 equally spaced U-shaped cannulas inserted just below the gradient surface. Each cannula acts to deposit a tiny fraction of the total sample onto the gradient. In the same device, 92 conically shaped and tangential bores in the flat baseplate of the chamber are used to collect the separated zones at the end of the run (Fig. 11-4).

Fig. 11-4 Multiple port loading and unloading sta-put device of Zeiller and co-workers (see Zeiller et al., 1976). Courtesy of Dr. G. Pascher.

In a sta-put device used by the author and his colleagues (Fig. 11-5), the conical upper end cap acts like a piston in the cylindrical separation chamber and can be caused to slide upward or downward by introduction or removal of gradient (Sheeler and Doolittle, 1980). The operation of the sta-put is depicted in Fig. 11-6. With the upper conical end cap in its fully depressed position, cushion is pumped into the lower end cap until it emerges through the port in the upper end cap (Fig. 11-6A). At this point the gradient is pumped, dense end first, into the sta-put chamber through the upper cone. With the lower port closed (Fig. 11-6B), the upper cone is caused to rise in the chamber. Once the entire density gradient has been introduced, the sample is gently drawn onto its surface by removing some of the cushion (Fig. 11-6C). In a like manner, an overlay is drawn into the chamber to displace the sample clear of the upper cone (Fig. 11-6D). After the particles in the sample have been separated (Fig. 11-6E), the gradient (and the particle zones) is (are) collected in either of three ways: (1) the gradient is drained out of the sta-put through the lower cone (Fig.

11-6F1); (2) the gradient is displaced up and out of the sta-put through the upper cone by using additional cushion (Fig. 11-6F2); or (3) the lower port is closed and the gradient drawn out through the upper port, thereby causing the piston to descend through the chamber to "meet" each particle zone (Fig. 11-6F3). The last approach provides the advantage of not requiring that the separated zones move (either upward or downward) through the apparatus until they enter the conical piston.

Reorienting Sta-Put Devices   To obviate the need for conical end caps, MacDonald and Miller (1970) introduced the idea of slightly tilting a rectangular chamber on one of its edges during the loading and unloading stages of operation. In this way, the incoming or outgoing liquid stream can be directed to one corner (which is fitted with a port), where the converging chamber walls act much like a cone. A similar approach has been taken by a number of other investigators (e.g., Bont and Hilgers, 1977). In previously unreported studies, the author and M. H. Doolittle

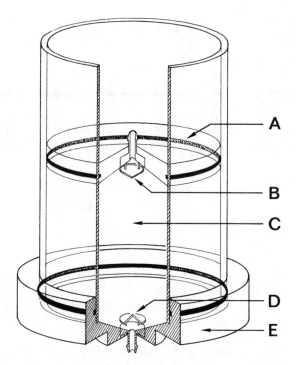

Fig. 11-5   Piston-type sta-put device used by the author (see text for description): A, floating end cap (piston); B, upper baffle; C, separation chamber; D, lower baffle; E, fixed lower end cap.

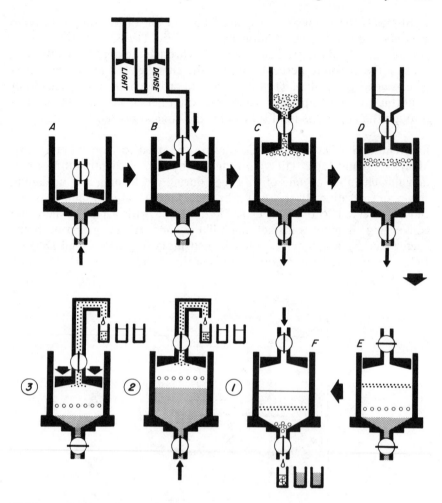

Fig. 11-6 Stages of operation of the piston-type sta-put device (see text for explanation).

have substituted a shallow, cylindrical chamber for a rectangular one (Fig. 11-7). Ports were fitted at the lowermost and uppermost circumferential edges of the chamber for delivery and/or removal of the density gradient and the particle zones.

Sta-put devices employing the tilting feature are sometimes called *reorienting sta-puts* or *reo-puts,* although in this instance it is the chamber that is being reoriented—not the density gradient. At least in theory, reorienting sta-puts offer some interesting advantages over stationary cylinders with conical caps, and these are shown in Fig. 11-8, which depicts the operating scheme of a typical device. The density gradient is pumped (light end first) into the lower port of the inclined chamber followed by the cushion (Fig. 11-8A). With the chamber full, the sample and the overlay are drawn into the chamber by draining some of the cushion through the lower port (Fig. 11-8B). At this point the chamber is slowly and smoothly reoriented into the horizontal position. During reorientation, the sample is slowly spread over the increasing surface area of the light end of the gradient and in this manner forms a much narrower band than when the chamber was inclined (Fig. 11-8C). The particles are then allowed to sediment through the gradient to form a series of zones (Fig. 11-8D). With the separation complete, the chamber is again slowly tilted to its inclined position (Fig. 11-8E) Although the volume of particle-free gradient between successive zones is not altered by reorientation of the chamber, the vertical distance between the zones is increased (compare Figs. 11-8D and 11-8E). Consequently, during collection of the gradient (Fig. 11-8F), there is less chance of cross contamination of the separated particles. In Fig. 11-8 the density gradient and particles are collected through the lower port; however, displacement through the upper port using cushion is an alternative procedure. Tilting the chamber of a reorienting sta-put device can be achieved by use of mechanical linkages and gears (as in Fig. 11-7) or by simple hydraulic mechanisms (Bont and Hilgers, 1977).

Despite the diversity of shapes and forms of reorienting and stationary sta-put devices, no comprehensive comparison of the effectiveness of the

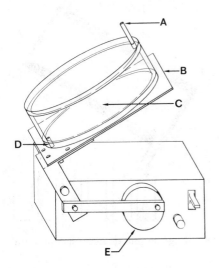

Fig. 11-7 A reorienting sta-put device (i.e., "reoput"): *A*, upper loading–unloading port; *B*, chamber platform; *C*, separation chamber; *D*, lower loading–unloading port; *E*, motor and control unit for reorienting chamber.

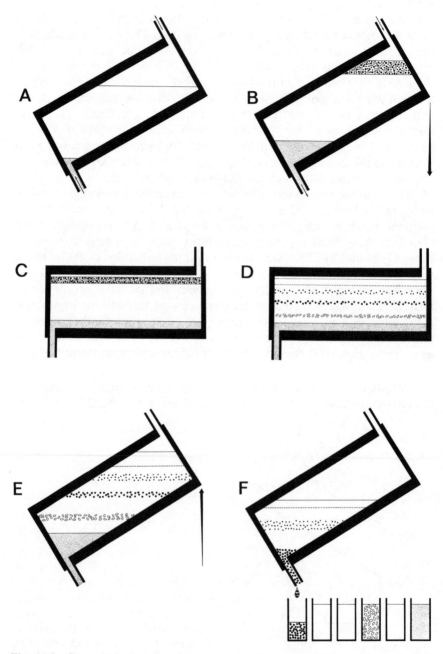

Fig. 11-8 Stages of operation of the reorienting sta-put device (see text for explanation).

various designs has been made. Using polystyrene microspheres of discrete size classes as a model test system, the author and M. H. Doolittle compared resolution in a reorienting sta-put and a conventional cylindrical chamber fitted with conical end caps; no differences in effectiveness were noted (unpublished observations). Zeiller et al. (1976), using a similar test system, compared a number of configurations in which stationary chambers were loaded and unloaded through various numbers and positions of entry and exit ports. They, too, found no appreciable differences in the resolution attainable.

CHOICE OF GRADIENTS FOR UNIT GRAVITY SEPARATIONS

The choice of density gradients to be used for unity gravity separations is somewhat more restricted than for centrifugal separations. The principal reason for this is that at $1 \times g$, liquids of high density or viscosity so greatly attenuate sedimentation rates that the time required for particles to fall even modest distances can be inordinately long. Since unit gravity devices are generally used for rate separations, the density of the gradient is normally quite low; however, a number of gradient solutes produce high viscosities, even at low density (e.g., Ficoll; see Fig. 4-20). This creates a special problem where separations of cells are concerned, since the high-viscosity solutes are also those that have the smallest osmotic effects on cells (see Fig. 4-19). Therefore planning the "best" gradient for a given cell separation at unit gravity is often a compromise between practical and ideal considerations.

STA-PUT SEPARATIONS VERSUS CENTRIFUGAL SEPARATIONS

With respect to the separation of cells or other large particles by sedimentation rate differences, the main alternatives to unit gravity approaches are elutriation and density gradient centrifugation. Elutriation offers the special advantage of maintaining the cells in a constant and physiologically optimal environment throughout the separation procedure (see Chapter 10), since no density gradient is required; however, the total number of cells that can be handled in a single operation is somewhat limited.

Several different forms of density gradient centrifugation have been used to separate whole cells. Isopycnic separations are readily achievable when density differences exist. However, since isopycnic separations require much higher gradient densities than do rate separations, the deleterious effects on cells of hypertonic or hyperosmotic conditions limit the applicability of this approach to especially resistant cells. The much greater variation that exists in cell sizes than cell densities has already been noted. The sedimentation coefficients of most cells are so high that rate separations in shallow, low-density (and isoosmotic) gradients present a special problem because the time of centrifugation may have to be limited to several seconds or minutes. The sedimentation rate of the cells can be reduced by operating at the lowest angular velocity commensurate with complete gradient reorientation and/or by using rotors of low minimum and maximum radii. For example, swinging-bucket rotors may be modified to bring the buckets closer to the axis of rotation. However, the numbers of cells that can be separated is usually restricted because of the small capacities of most buckets.

As noted in Chapter 6, sample capacity is greatly increased in *zonal* rotors so that quantitative separations of cells comparable to those attainable in multiliter unit gravity systems should be possible. Among dynamically unloaded zonal rotors, A types have been used successfully for the separation of certain cells because these rotors can be loaded, operated, and then unloaded at low speeds. Reorienting gradients zonal rotors have also been effective for rate separations of cells in shallow, low-density gradients because they not only offer high sample capacity, but are unloaded at $1 \times g$. Nearly all the particle sedimentation that occurs can be confined to the time at speed, and this can be limited to just a few seconds.

REFERENCES AND RELATED READING

Books
Catsimpoolas, N., Ed. *Methods of Cell Separation*, Vol. 1. Plenum, New York, 1977.

Catsimpoolas, N., Ed. *Methods of Cell Separation*, Vol. 2. Plenum, New York, 1979.

Cutts, J. H. *Cell Separation Methods in Hematology*. Academic, New York, 1970.

Articles and Reviews
Bont, W. S., and Hilgers, J. H. M. (1977) Rapid separation of cells at unit gravity. *Preparative Biochem.*, **7**, 45.

Brubaker, L. H., and Evans, W. H. (1969) Separation of granulocytes, monocytes, lymphocytes, erythrocytes, and platelets from human blood and relative tagging with diisopropylfluorophosphate (DFP). *J. Lab. Clin. Med.*, **73**, 1036.

MacDonald, H. R., and Miller, R. G. (1970) Synchronization of mouse L-cells by a velocity sedimentation technique. *Biophys. J.*, **10**, 834.

Meistrich, M. L. (1977) Separation of spermatogenic cells and nuclei from rodent testes. In *Methods in Cell Biology*, Vol. XV. (D. M. Prescott, Ed.) Academic, New York.

Mel, H. C. (1963) Sedimentation properties of nucleated and non-nucleated cells in normal rat bone marrow. *Nature*, **200**, 423.

Mel, H. C. (1964) Stable-flow free boundary (STAFLO) migration and fractionation of cell mixtures. I. Apparatus and hydrodynamic feedback principles. *J. Theoret. Biol.*, **6**, 159.

Mel, H. C., and Mohandas, N. (1979) Physical separation and characterization of reticulocytes and other cell fractions from rat bone marrow and the 1 g ministaflo. In *Methods of Cell Separation*, N. Catsimpoolas, Ed. Plenum, New York.

Miller, R. G. (1973) Separation of cells by velocity sedimentation. In *New Techniques in Biophysics and Cell Biology*, R. H. Pain and B. J. Smith, Eds. Wiley-Interscience, London.

Miller, R. G., and Phillips, R. A. (1969) Separation of cells by velocity sedimentation. *J. Cell. Physiol.*, **73**, 191.

Peterson, E. A., and Evans, W. H. (1967) Separation of bone marrow cells by sedimentation at unit gravity. *Nature*, **214**, 824.

Phillips, R. A., and Miller, R. G. (1970) Antibody-producing cells: Analysis and purification by velocity sedimentation. *Tissue Kinet.*, **3**, 263.

Pretlow, T. G., Weir, E. E., and Zettergren, J. G. (1975) Problems connected with the separation of different kinds of cells. In *International Review of Experimental Pathology*, Vol. 14. Academic, San Francisco.

Sheeler, P., and Doolittle, M. H. (1980) Separation of mammalian cells by velocity sedimentation. *Amer. Lab.*, **12** (4), 49.

Tait, J. F., Tait, S. A. S., Gould, R. P., and Mee, M. S. R. (1974) The properties of adrenal zona glomerulosa cells after purification by gravitational sedimentation. *Proc. Roy. Soc., London, B*, **185**, 375.

Tulp, A., and Bont, W. S. (1975) An improved method for the separation of cells by sedimentation at unit gravity. *Anal. Biochem.*, **67**, 11.

Zeiller, K., and Hansen, E. (1978) Characterization of rat bone marrow lymphoid cells. I. A study of the distribution of parameters of sedimentation velocity, volume and electrophoretic mobility. *J. Histochem. Cytochem.*, **26**, 369.

Zeiller, K., and Hansen, E. (1979) Characterization of rat bone marrow cells. II. Analysis of surface antigens in small lymphocytes with particular reference to thymus antigen-carrying cells. *Cell. Immunol.*, **44**, 381.

Zeiller, K., Hansen, E., Leihener, D., Pascher, G., and Wirth, H. (1976) Analysis of velocity sedimentation techniques in cell separation. *Hoppe-Seyler's Z. Physiol. Chem.*, **357**, 1309.

Appendixes

There are three appendixes: A, a series of tables listing the specifications of the most frequently used ultraspeed and superspeed centrifuge rotors; B, a series of tables listing the physical properties of gradient solutions; and C, a list of the manufacturers of centrifuges, rotors, and other instruments and materials used in centrifugation.

The tables of values presented in Appendix B are for aqueous solutions of the various gradient solutes and for solutes dissolved in 0.15 M NaCl. Although the values listed for the aqueous solutions were obtained from various sources (cited in the footnotes to each table), the values presented for gradient solutes dissolved in saline solution were obtained experimentally by the author and Etan Milgrom; these data have not been published previously.

The temperatures at which the measurements were made for gradient solutes dissolved in saline solution were chosen because they closely approximate the temperature at which typical density gradient separations are performed (i.e., at 4°C) and the temperature at which collected fractions are subsequently analyzed (i.e., at room temperature or 22°C).

Tables of Rotor
Specifications

Table A-1 Ultraspeed Fixed-Angle Rotors

Manufacturer	Model Number	rpm (max)	RCF			Angle (°)	Positions	Position Volume (ml)	Total Capacity (ml)	K Factor
			max	ave	min					
DuPont/Sorvall	T875	75,000	547,300	418,500	290,300	23.5	8	12.5	100	28
	T865	65,000	429,500	305,300	181,200	23.5	8	36	288	52
	T865.1	65,000	411,100	314,300	218,000	23.5	8	12.5	100	38
	A641	41,000	186,500	125,100	63,700	22.5	6	100	600	162
	A841	41,000	170,900	121,500	72,100	23.5	8	36	288	130
Beckman/Spinco	80Ti	80,000	602,000	447,800	293,600	25.5	8	13.5	108	28
	75Ti	75,000	503,500	368,200	232,900	25.5	8	13.5	108	35
	70Ti	70,000	504,000	356,100	213,700	23	8	38.5	308	44
	70.1Ti	70,000	450,000	334,700	219,300	24	12	13.5	162	37
	Type 65	65,000	368,800	269,500	174,900	23.5	8	13.5	108	45
	60Ti	60,000	362,600	253,800	149,100	23.5	8	38.5	308	63
	55.2Ti	55,000	340,000	249,200	158,400	24	10	38.5	385	64
	50.2Ti	50,000	303,000	227,300	148,700	24	12	38.5	462	72
	50Ti	50,000	226,600	165,100	106,300	26	12	13.5	162	76
	50.3Ti	50,000	223,800	178,800	133,900	20	18	6.5	117	52
	50	50,000	198,700	151,100	103,500	20	10	10	100	66
	45Ti	45,000	235,800	167,000	83,900	24	6	94	564	129
	LP42Ti	42,000	222,000	215,200	208,500	30	72	0.175	12.6	9
	42Ti	42,000	195,000	136,000	76,800	30	8	38.5	308	134
	Type 40	40,000	145,000	105,600	68,000	20	12	13.5	162	120
	Type 40.3	40,000	143,200	114,600	85,900	20	18	6.5	117	81
	Type 40.2	40,000	143,100	102,000	62,600	40	12	6.5	78	131
	Type 35	35,000	142,800	96,100	48,100	25	6	94	564	225

Table A-1 (*Continued*)

Manufacturer	Model Number	RCF				Angle (°)	Positions	Position Volume (ml)	Total Capacity (ml)	K Factor
		rpm (max)	max	ave	min					
	Type 30	30,000	106,000	78,500	50,300	26	12	38.5	462	209
	Type 30.2	30,000	94,500	79,400	63,300	14	20	10.5	210	113
	Type 25	25,000	92,300	85,800	79,200	25	100	1	100	62
	Type 21	21,000	59,200	44,400	29,600	18	10	94	940	398
	Type 19	19,000	53,700	35,900	17,800	25	6	250	1,500	776
	Type 15	15,000	35,800	23,000	10,200	15	4	500	2,000	1,405
Damon/IEC	A321	60,000	321,400	214,900	108,400	35	8	12	96	76
	A269	55,000	269,100	205,100	141,200	20	10	12	120	53
	A237	50,000	237,200	165,500	93,800	23	8	40	320	92
	A211	45,000	211,200	140,000	69,600	30	8	40	320	137
	A192	40,000	192,000	129,000	66,000	33	8	50	400	168
	A170	40,000	170,300	113,300	56,200	20	6	75	450	176
	A168	40,000	168,600	137,500	106,300	14	20	12	240	60
	A110	30,000	109,800	81,900	53,900	26	12	40	480	199
	A54	20,000	54,800	35,400	16,100	20	6	250	1,500	767
	A28	14,000	28,500	14,400	3,200	20	4	500	2,000	2,854

MSE

8 × 14Ti	75,000	511,000	365,900	220,800	29	8	14	112	38
10 × 10Ti	72,000	499,400	374,500	249,700	35	10	10	100	34
10 × 10Ti	65,000	407,000	291,500	176,000	35	10	10	100	42
8 × 35Ti	60,000	379,200	262,200	145,200	21	8	35	260	68
8 × 25Ti	60,000	379,200	278,400	177,500	30	8	25	200	54
8 × 14Al	65,000	368,500	266,900	165,300	25.5	8	14	112	46
8 × 35Al	50,000	263,000	181,000	100,700	21	8	35	260	98
8 × 25Al	50,000	263,000	193,000	123,100	30	8	25	200	78
10 × 10Al	50,000	199,200	154,200	109,100	20	10	10	100	61
8 × 50Al	40,000	195,000	138,600	82,300	30	8	50	400	137
6 × 100Al	35,000	154,000	104,500	55,000	25	6	100	600	213
8 × 10Al	40,000	142,400	104,300	66,200	35	8	10	80	127
10 × 100Al	25,000	88,500	66,200	43,900	18	10	100	1,000	284
6 × 300Al	21,000	75,000	51,900	28,900	25	6	300	1,800	547
4 × 500Al	14,000	32,200	21,000	9,000	18	4	500	2,000	1,650

Table A-2 Ultraspeed Swinging-Bucket Rotors

Manufacturer	Model Number	rpm (max)	RCF max	RCF ave	RCF min	Positions	Position Volume (ml)	Total Capacity (ml)	K Factor
DuPont/Sorval	AH650	50,000	299,900	233,500	167,000	6	5	30	59
	AH627-17	27,000	135,100	94,900	55,000	6	17	102	313
	AH627-36	27,000	131,100	96,700	62,400	6	36	216	258
Beckman/Spinco	SW65Ti	65,000	420,400	302,300	194,600	3	5	15	46
	SW60Ti	60,000	484,200	369,100	254,000	6	4.4	26.4	45
	SW55Ti	55,000	369,000	287,700	206,500	6	5	30	49
	SW50.1	50,000	300,000	233,400	166,900	6	4.4	26.4	59
	SW41Ti	41,000	286,200	205,400	124,600	6	13.2	79.2	125
	SW40Ti	40,000	285,000	202,600	119,000	6	14	84	137
	SW28.1	28,000	150,000	107,100	64,200	6	17	102	276
	SW28	28,000	141,000	103,600	66,100	6	38.5	231	245
	SW27.1	27,000	135,200	91,700	53,200	6	17	102	325
	SW27	27,000	131,400	96,400	61,400	6	38.5	231	265
	SW25.2	25,000	106,400	76,500	46,500	3	60	180	335
	SW25.1	25,000	90,300	64,800	39,300	3	34	102	338

Damon/IEC								
SB405	60,000	405,900	305,400	205,000	6	4.2	25.2	48
SB283	41,000	283,200	193,500	103,700	6	14	84	151
SB110	25,000	110,000	77,700	45,500	6	40	240	355
MSE								
6 × 4.2	60,000	500,000	354,500	209,000	6	4.2	25.2	62
3 × 6.5	60,000	420,000	296,000	172,000	3	6.5	19.5	63
3 × 5	50,000	300,000	281,100	131,100	3	5	15	84
6 × 5.5	45,000	242,700	185,800	128,900	6	5.5	33	80
6 × 14	40,000	284,000	198,800	113,500	6	14	84	145
6 × 16.5	30,000	159,000	110,300	61,600	6	16.5	99	267
3 × 25	30,000	129,800	94,600	59,400	3	25	75	220
6 × 38	25,000	115,300	85,800	56,400	6	38	228	288
3 × 70	23,500	100,000	69,800	39,500	3	65	195	425

Table A-3 Ultraspeed Vertical Tube Rotors

Manufacturer	Model Number	rpm (max)	RCF			Positions	Position Volume (ml)	Total Capacity (ml)	K Factor
			max	ave	min				
DuPont/Sorvall	TV 865	65,000	400,600	369,400	338,300	8	5	40	10
	TV 865B	65,000	400,000	340,600	281,500	8	17	136	21
	TV 850	50,000	236,500	201,000	165,400	8	35	280	36
Beckman/Spinco	VTi 80	80,000	510,000	462,300	414,600	8	5.1	40.8	8
	VTi 65	65,000	402,000	371,000	340,300	8	5.1	40.8	10
	VTi 50	50,000	241,200	207,500	173,800	8	39	312	33
	VAl 26	26,000	70,000	60,200	50,400	8	39	312	123
MSE	VWR 65	65,000	401,000	345,000	289,000	8	5	40	8
	VWR 50	50,000	240,600	207,000	173,400	8	35	280	34

Table A-4 Ultraspeed Zonal Rotors

Manufacturer	Model Number	Anderson Designation	Type	Maximum rpm	Maximum RCF	Capacity (ml)	Path (cm)	Comment
DuPont/Sorvall	TZ-28Ti	—	Reograd	28,000	83,500	1,350	5.9	
Beckman/Spinco	Z-60Ti	—	Dynamic	60,000	256,000	330	5.2	
	Al-14	B-XIV	Dynamic	35,000	91,300	650	5.4	
	Ti-14	B-XIV	Dynamic	48,000	171,800	650	5.4	
	Al-15	B-XV	Dynamic	22,000	48,000	1,665	7.6	
	Ti-15	B-XV	Dynamic	31,000	102,000	1,665	7.6	
	B-29	B-XXIX	Dynamic	—[a]	—[a]	525	—[a]	Inserts for Al-14 and Ti-14
	B-29	B-XXX	Dynamic	—[a]	—[a]	1,350	—[a]	Inserts for Al-15 and Ti-15
Damon/IEC	B-14Ti	B-XIV	Dynamic	50,000	186,500	659	5.1	
	B-15Ti	B-XV	Dynamic	35,000	121,800	1,674	7.4	
	B-29Ti	B-XXIX	Dynamic	35,000	121,800	1,480	7.4	
	B-30Ti	B-XXX	Dynamic	50,000	186,500	570	5.1	
MSE	B-14Al	B-XIV	Dynamic	35,000	91,300	650	5.4	
	B-14Ti	B-XIV	Dynamic	47,000	165,000	650	5.4	
	B-15Al	B-XV	Dynamic	25,000	62,000	1,670	7.6	
	B-15Ti	B-XV	Dynamic	35,000	121,800	1,670	7.6	
	B-29	B-XXIX	Dynamic	—[a]		1,420	—[a]	Inserts for B-15Al, and B-15Ti

[a] Although the B-29 and B-30 inserts reduce the total capacity of the rotor, they do not appreciably affect the maximum rpm (speed), maximum RCF, or the path length.

Table A-5 Superspeed Fixed-Angle Rotors

Manufacturer	Model Number	rpm (max)	RCF			Angle (°)	Positions	Position Volume (ml)	Total Capacity (ml)	K Factor
			max	ave	min					
DuPont/Sorvall	SS-34	20,000	48,200	36,900	25,600	34	8	50	400	401
	SM-24	20,000	49,300	40,300	31,300	28	24	17.5	420	386
	SE-12	20,000	41,500	32,600	23,600	40	12	15	180	357
	SA-600	16,500	39,400	31,800	24,100	34	12	50	600	455
	GSA	13,000	27,600	22,500	17,500	28	6	315	1,890	680
	GS-3	9,000	13,700	11,600	9,500	25	6	500	3,000	1,153
Beckman/Spinco	JA-20.1	20,000	51,500	40,000	28,500	23	32	15	480	468
	JA-21	21,000	50,300	35,300	20,200	40	18	10	180	523
	JA-20	20,000	48,300	31,300	14,300	34	8	50	400	770
	JA-17	17,000	39,700	28,900	18,100	25	14	50	700	688
	JA-14	14,000	30,000	19,200	8,300	25	6	250	1,500	1,655
	JA-10	10,000	17,700	11,000	4,300	25	6	500	3,000	3,607
Damon/IEC	870	19,500	45,500	33,500	21,500	33	8	50	400	498
	873	20,000	39,600	31,000	22,400	40	12	14	168	359
	874	19,500	39,200	27,800	16,400	33	24	13	312	581
	872	14,500	28,900	19,000	9,000	20	6	250	1,500	2,949

Table A-6 Superspeed Swinging-Bucket Rotors

Manufacturer	Model Number	rpm (max)	RCF max	RCF ave	RCF min	Positions	Position Volume (ml)	Total Capacity (ml)	K Factor
DuPont/Sorvall	HB-4	13,000	27,600	18,300	9,000	4	50	200	1,682
	HS-4	7,000	9,400	6,700	4,000	4	250	1,000	4,495
Beckman/Spinco	JS-13	13,000	26,900	17,200	7,600	4	50	200	1,897
	JS 7.5	7,500	10,400	5,400	1,600	4	250	1,000	8,311
Damon/IEC	940	12,000	20,800	13,700	6,500	4	40	160	2,030
	947	5,000	5,000	3,200	1,300	4	150	600	13,560

Table A-7 Superspeed Vertical Tube Rotors

Manufacturer	Model Number	rpm (max)	RCF max	RCF ave	RCF min	Positions	Position Volume (ml)	Total Capacity (ml)	K Factor
DuPont/Sorvall	SV-288	20,000	40,300	34,600	29,000	8	36	288	210
	SV-80	19,000	41,000	38,300	35,700	16	5	80	97
Beckman/Spinco	JV-20	20,000	41,400	35,800	30,100	8	36	288	203

Table A-8 Superspeed Zonal Rotors

Manufacturer	Model Number	Anderson Designation	Type	Maximum rpm	Maximum RCF	Capacity (ml)	Path (cm)	Comment
DuPont/Sorvall	SZ-14	—	Reograd	19,500	40,500	1,350	5.9	
	TZ-28Ti	—	Reograd	20,000	42,600	1,350	5.9	
Beckman/Spinco	JCF-Z	—	Reograd	20,000	40,000	1,750	6.9	
	JCF-Z	—	Dynamic	20,000	40,000	1,900	6.9	
Damon/IEC	Z-15	—	Dynamic	8,000	9,100	780	9.5	Transparent upper and lower end plates
MSE	A-Type	A-XII	Dynamic	5,000	5,000	1,300	13.6	Transparent upper and lower end plates
	HS	—	Dynamic	10,000	12,800	695	7.6	Transparent upper and lower end plates

Physical Properties of Gradient Solutions

Table B-1 Physical Properties of Aqueous CsCl Solutions at 20°C

Concentration		Refractive Index	Density (gm/ml)	Viscosity (cP)
Percent w/w	Percent w/v			
0.00	0.000	1.3330	0.9988	1.004
1.00	1.005	1.3333	1.0059	0.996
2.00	2.027	1.3341	1.0137	0.992
4.00	4.119	1.3357	1.0297	0.984
6.00	6.277	1.3373	1.0461	0.976
8.00	8.504	1.3389	1.0630	0.970
10.0	10.804	1.3406	1.0804	0.963
12.0	13.180	1.3425	1.0983	0.959
14.0	15.635	1.3443	1.1168	0.953
16.0	18.173	1.3461	1.1358	0.948
18.0	20.799	1.3480	1.1555	0.942
20.0	23.516	1.3500	1.1758	0.939
22.0	26.330	1.3520	1.1968	0.934
24.0	29.244	1.3541	1.2185	0.930
26.0	32.263	1.3563	1.2409	0.925
28.0	35.394	1.3586	1.2641	0.920
30.0	38.646	1.3610	1.2882	0.918
35.0	47.327	1.3673	1.3522	0.918
40.0	56.900	1.3725	1.4225	0.928
45.0	67.495	1.3810	1.4999	0.947
50.0	79.290	1.3887	1.5858	0.984
55.0	92.477	1.3975	1.6814	1.038
60.0	107.32	1.4075	1.7886	1.128
65.0	124.13	1.4185	1.9097	1.288

Source: Griffith, O. M., *Techniques of Preparative, Zonal, and Continuous Flow Ultracentrifugation,* Applications Research Department, Spinco Division, Beckman Instruments, Inc., Palo Alto, California.

Table B-2 Physical Properties of CsCl in Saline at 4°C

Concentration				
Percent w/w	Percent w/v	Refractive Index	Density (gm/ml)	Viscosity (cP)
0	0	1.3361	1.0057	1.58
0.99	1	1.3368	1.0132	1.56
1.96	2	1.3376	1.0206	1.54
2.92	3	1.3384	1.0281	1.52
3.86	4	1.3391	1.0355	1.50
4.79	5	1.3398	1.0430	1.49
5.71	6	1.3406	1.0504	1.48
6.62	7	1.3414	1.0579	1.47
7.51	8	1.3421	1.0653	1.46
8.39	9	1.3429	1.0728	1.45
9.26	10	1.3436	1.0803	1.44
10.11	11	1.3444	1.0877	1.43
10.95	12	1.3451	1.0955	1.42
11.79	13	1.3459	1.1026	1.42
12.61	14	1.3466	1.1101	1.41
13.42	15	1.3474	1.1175	1.40
14.22	16	1.3481	1.1250	1.40
15.01	17	1.3489	1.1324	1.39
15.79	18	1.3496	1.1399	1.38
16.56	19	1.3504	1.1473	1.37
17.32	20	1.3511	1.1548	1.37
18.07	21	1.3519	1.1622	1.37
18.81	22	1.3526	1.1697	1.36
19.54	23	1.3534	1.1772	1.36
20.26	24	1.3541	1.1846	1.36
20.97	25	1.3549	1.1921	1.35
21.68	26	1.3556	1.1995	1.35
22.37	27	1.3564	1.2070	1.34
23.06	28	1.3571	1.2144	1.34
23.73	29	1.3579	1.2219	1.33
24.40	30	1.3586	1.2293	1.33
25.06	31	1.3594	1.2368	1.33

Table B-2 *(Continued)*

Concentration		Refractive Index	Density (gm/ml)	Viscosity (cP)
Percent w/w	Percent w/v			
25.72	32	1.3601	1.2443	1.32
26.36	33	1.3609	1.2517	1.32
27.00	34	1.3616	1.2592	1.32
27.63	35	1.3624	1.2666	1.31
28.26	36	1.3632	1.2741	1.31
28.87	37	1.3639	1.2815	1.31
29.48	38	1.3647	1.2890	1.30
30.08	39	1.3654	1.2964	1.30
30.68	40	1.3662	1.3039	1.30
31.27	41	1.3669	1.3113	1.30
31.85	42	1.3677	1.3188	1.29
32.42	43	1.3684	1.3263	1.29
32.99	44	1.3692	1.3337	1.29
33.55	45	1.3699	1.3412	1.29
34.11	46	1.3707	1.3486	1.29
34.66	47	1.3714	1.3561	1.28
35.20	48	1.3722	1.3635	1.28
35.74	49	1.3729	1.3710	1.28
36.27	50	1.3737	1.3784	1.28
36.80	51	1.3744	1.3859	1.28
37.31	52	1.3752	1.3936	1.28
37.84	53	1.3759	1.4008	1.28
38.34	54	1.3767	1.4083	1.27
38.85	55	1.3774	1.4157	1.27
39.35	56	1.3782	1.4232	1.27
39.84	57	1.3789	1.4306	1.27
40.33	58	1.3797	1.4381	1.27
40.82	59	1.3804	1.4455	1.27
41.29	60	1.3812	1.4530	1.27

Source: Milgrom, E., and Sheeler, P. (previously unpublished).

Table B-3 Physical Properties of CsCl in Saline at 22°C

Concentration		Refractive	Density	Viscosity
Percent w/w	Percent w/v	Index	(gm/ml)	(cP)
0	0	1.3349	1.0033	0.96
0.99	1	1.3356	1.0107	0.96
1.96	2	1.3363	1.0182	0.95
2.93	3	1.3371	1.0256	0.95
3.87	4	1.3378	1.0330	0.95
4.81	5	1.3385	1.0404	0.95
5.73	6	1.3392	1.0479	0.95
6.63	7	1.3399	1.0553	0.94
7.53	8	1.3406	1.0627	0.94
8.41	9	1.3414	1.0702	0.94
9.28	10	1.3421	1.0776	0.94
10.14	11	1.3428	1.0850	0.94
10.98	12	1.3435	1.0924	0.94
11.82	13	1.3442	1.0999	0.93
12.64	14	1.3450	1.1073	0.93
13.46	15	1.3457	1.1147	0.93
14.26	16	1.3464	1.1222	0.93
15.05	17	1.3471	1.1296	0.93
15.83	18	1.3478	1.1370	0.93
16.60	19	1.3485	1.1444	0.93
17.36	20	1.3493	1.1519	0.93
18.11	21	1.3500	1.1593	0.92
18.86	22	1.3507	1.1667	0.92
19.59	23	1.3514	1.1742	0.92
20.31	24	1.3521	1.1816	0.92
21.03	25	1.3529	1.1890	0.92
21.73	26	1.3536	1.1964	0.92
22.43	27	1.3543	1.2039	0.92
23.12	28	1.3550	1.2113	0.92
23.80	29	1.3557	1.2187	0.91
24.47	30	1.3565	1.2261	0.91
25.13	31	1.3572	1.2336	0.91

Table B-3 (*Continued*)

Concentration		Refractive Index	Density (gm/ml)	Viscosity (cP)
Percent w/w	Percent w/v			
25.79	32	1.3579	1.2410	0.91
26.43	33	1.3586	1.2484	0.91
27.07	34	1.3593	1.2559	0.91
27.71	35	1.3600	1.2633	0.91
28.33	36	1.3608	1.2707	0.91
28.95	37	1.3615	1.2781	0.91
29.56	38	1.3622	1.2856	0.91
30.16	39	1.3629	1.2930	0.91
30.76	40	1.3636	1.3004	0.90
31.35	41	1.3644	1.3079	0.90
31.93	42	1.3651	1.3153	0.90
32.51	43	1.3658	1.3227	0.90
33.08	44	1.3665	1.3301	0.90
33.64	45	1.3672	1.3376	0.90
34.20	46	1.3679	1.3450	0.90
34.75	47	1.3687	1.3524	0.90
35.30	48	1.3694	1.3599	0.90
35.84	49	1.3701	1.3673	0.90
36.37	50	1.3708	1.3747	0.90
36.90	51	1.3715	1.3821	0.90
37.42	52	1.3723	1.3896	0.90
37.94	53	1.3730	1.3970	0.90
38.45	54	1.3737	1.4044	0.90
38.95	55	1.3744	1.4119	0.90
39.46	56	1.3751	1.4193	0.90
39.95	57	1.3759	1.4267	0.90
40.44	58	1.3766	1.4341	0.90
40.93	59	1.3773	1.4416	0.90
41.41	60	1.3780	1.4490	0.91

Source: Milgrom, E., and Sheeler, P. (previously unpublished).

Table B-4 Physical Properties of Aqueous Sucrose Solutions at 4°C

Concentration		Refractive Index	Density (gm/ml)	Viscosity (cP)
Percent w/w	Percent w/v			
0.000	0.000	1.3346	1.0004	1.564
1.000	1.002	1.3361	1.0043	1.608
2.000	2.012	1.3376	1.0082	1.654
3.000	3.030	1.3392	1.0122	1.703
4.000	4.056	1.3408	1.0161	1.754
5.000	5.089	1.3424	1.0202	1.810
6.000	6.131	1.3439	1.0242	1.868
7.000	7.181	1.3454	1.0283	1.930
8.000	8.240	1.3468	1.0324	1.996
9.000	9.306	1.3483	1.0366	2.067
10.00	10.38	1.3498	1.0407	2.142
11.00	11.47	1.3516	1.0450	2.222
12.00	12.56	1.3532	1.0492	2.307
13.00	13.66	1.3548	1.0535	2.398
14.00	14.77	1.3564	1.0578	2.496
15.00	15.89	1.3580	1.0622	2.600
16.00	17.02	1.3595	1.0666	2.712
17.00	18.15	1.3610	1.0710	2.831
18.00	19.30	1.3626	1.0755	2.960
19.00	20.45	1.3642	1.0800	3.099
20.00	21.62	1.3658	1.0845	3.248
21.00	22.79	1.3676	1.0891	3.408
22.00	23.98	1.3694	1.0937	3.582
23.00	25.17	1.3711	1.0983	3.770
24.00	26.38	1.3725	1.1030	3.973
25.00	27.59	1.3741	1.1077	4.194
26.00	28.81	1.3758	1.1124	4.434
27.00	30.05	1.3776	1.1172	4.695
28.00	31.29	1.3794	1.1220	4.981
29.00	32.54	1.3811	1.1268	5.293
30.00	33.81	1.3828	1.1317	5.636
31.00	35.08	1.3847	1.1366	6.012
32.00	36.37	1.3866	1.1416	6.427
33.00	37.67	1.3884	1.1466	6.886
34.00	38.98	1.3902	1.1516	7.394

Table B-4 *(Continued)*

| Concentration | | Refractive | Density | Viscosity |
Percent w/w	Percent w/v	Index	(gm/ml)	(cP)
35.00	40.29	1.3921	1.1566	7.959
36.00	41.62	1.3939	1.1617	8.588
37.00	42.97	1.3958	1.1668	9.293
38.00	44.32	1.3977	1.1720	10.08
39.00	45.68	1.3999	1.1772	10.97
40.00	47.06	1.4021	1.1824	11.98
41.00	48.45	1.4039	1.1876	13.12
42.00	49.84	1.4058	1.1929	14.42
43.00	51.26	1.4079	1.1983	15.90
44.00	52.68	1.4100	1.2036	17.59
45.00	54.11	1.4121	1.2090	19.54
46.00	55.56	1.4142	1.2144	21.80
47.00	57.02	1.4163	1.2199	24.41
48.00	58.90	1.4183	1.2254	27.44
49.00	59.98	1.4205	1.2310	32
50.00	61.48	1.4228	1.2365	37
51.00	62.99	1.4249	1.2418	43
52.00	64.51	1.4272	1.2472	50
53.00	66.05	1.4294	1.2530	57
54.00	67.60	1.4316	1.2589	65
55.00	69.16	1.4338	1.2649	75
56.00	70.74	1.4360	1.2708	90
57.00	72.33	1.4381	1.2767	103
58.00	73.94	1.4402	1.2826	122
59.00	75.56	1.4425	1.2884	150
60.00	77.19	1.4448	1.2941	180
61.00	78.83	1.4471	1.3000	220
62.00	80.49	1.4494	1.3060	275
63.00	82.17	1.4517	1.3120	340
64.00	83.86	1.4540	1.3181	425
65.00	85.56	1.4563	1.3242	555
66.00	87.28	1.4586	1.3305	725

Source: National Bureau of Standards (USA) and *Principles of Refractometry,* Bausch and Lomb Analytical Systems Division, Rochester, New York.

Table B-5 Physical Properties of Aqueous Sucrose Solutions at 20°C

Concentration		Refractive Index	Density (gm/ml)	Viscosity (cP)
Percent w/w	Percent w/v			
0.000	0.000	1.3330	0.9988	1.004
1.000	1.002	1.3344	1.0026	1.030
2.000	2.012	1.3359	1.0064	1.057
3.000	3.030	1.3373	1.0102	1.086
4.000	4.056	1.3388	1.0140	1.116
5.000	5.089	1.3403	1.0179	1.148
6.000	6.131	1.3418	1.0219	1.181
7.000	7.181	1.3433	1.0258	1.217
8.000	8.240	1.3448	1.0298	1.255
9.000	9.306	1.3463	1.0339	1.294
10.00	10.38	1.3478	1.0380	1.337
11.00	11.47	1.3494	1.0421	1.382
12.00	12.56	1.3509	1.0462	1.429
13.00	13.66	1.3525	1.0504	1.480
14.00	14.77	1.3541	1.0546	1.534
15.00	15.89	1.3557	1.0588	1.592
16.00	17.02	1.3573	1.0631	1.653
17.00	18.15	1.3589	1.0675	1.719
18.00	19.30	1.3605	1.0718	1.789
19.00	20.45	1.3622	1.0762	1.864
20.00	21.62	1.3638	1.0806	1.945
21.00	22.79	1.3655	1.0851	2.031
22.00	23.98	1.3672	1.0896	2.124
23.00	25.17	1.3689	1.0941	2.224
24.00	26.38	1.3706	1.0987	2.331
25.00	27.59	1.3723	1.1033	2.447
26.00	28.81	1.3740	1.1079	2.572
27.00	30.05	1.3758	1.1126	2.708
28.00	31.29	1.3775	1.1173	2.854
29.00	32.54	1.3793	1.1221	3.013
30.00	33.81	1.3811	1.1268	3.186
31.00	35.08	1.3829	1.1316	3.374
32.00	36.37	1.3847	1.1365	3.579
33.00	37.67	1.3865	1.1414	3.803
34.00	38.98	1.3883	1.1463	4.048

Table B-5 (*Continued*)

| Concentration | | Refractive | Density | Viscosity |
Percent w/w	Percent w/v	Index	(gm/ml)	(cP)
35.00	40.29	1.3902	1.1513	4.318
36.00	41.62	1.3920	1.1563	4.614
37.00	42.97	1.3939	1.1613	4.941
38.00	44.32	1.3958	1.1663	5.303
39.00	45.68	1.3978	1.1714	5.705
40.00	47.06	1.3997	1.1766	6.153
41.00	48.45	1.4016	1.1817	6.653
42.00	49.84	1.4036	1.1870	7.213
43.00	51.26	1.4056	1.1922	7.843
44.00	52.68	1.4076	1.1975	8.553
45.00	54.11	1.4096	1.2028	9.357
46.00	55.56	1.4117	1.2081	10.27
47.00	57.02	1.4137	1.2135	11.31
48.00	58.90	1.4158	1.2189	12.50
49.00	59.98	1.4179	1.2243	13.84
50.00	61.48	1.4200	1.2298	15.42
51.00	62.99	1.4221	1.2353	17.30
52.00	64.51	1.4242	1.2408	19.31
53.00	66.05	1.4264	1.2465	21.80
54.00	67.60	1.4285	1.2522	24.66
55.00	69.16	1.4307	1.2578	28.05
56.00	70.74	1.4329	1.2635	32.09
57.00	72.33	1.4351	1.2693	36.90
58.00	73.94	1.4373	1.2750	42.72
59.00	75.56	1.4396	1.2808	49.80
60.00	77.19	1.4418	1.2867	58.40
61.00	78.83	1.4441	1.2926	69.08
62.00	80.49	1.4464	1.2985	82.31
63.00	82.17	1.4486	1.3044	98.95
64.00	83.86	1.4509	1.3104	119.9
65.00	85.56	1.4532	1.3164	146.9
66.00	87.28	1.4555	1.3226	181.8

Source: National Bureau of Standards (USA) and *Principles of Refractometry,* Bausch and Lomb Analytical Systems Division, Rochester, New York.

Table B-6 Physical Properties of Aqueous Sucrose Solutions at 22°C

Concentration		Refractive	Density	Viscosity
Percent w/w	Percent w/v	Index	(gm/ml)	(cP)
0.000	0.000	1.3328	0.9984	0.956
1.000	1.002	1.3342	1.0022	0.981
2.000	2.012	1.3357	1.0059	1.006
3.000	3.030	1.3371	1.0097	1.033
4.000	4.056	1.3386	1.0136	1.062
5.000	5.089	1.3401	1.0175	1.092
6.000	6.131	1.3416	1.0214	1.124
7.000	7.181	1.3431	1.0254	1.157
8.000	8.240	1.3446	1.0293	1.192
9.000	9.306	1.3461	1.0334	1.230
10.00	10.38	1.3476	1.0374	1.270
11.00	11.47	1.3492	1.0415	1.312
12.00	12.56	1.3507	1.0457	1.356
13.00	13.66	1.3523	1.0498	1.404
14.00	14.77	1.3539	1.0540	1.455
15.00	15.89	1.3555	1.0583	1.509
16.00	17.02	1.3571	1.0626	1.566
17.00	18.15	1.3587	1.0669	1.628
18.00	19.30	1.3603	1.0712	1.693
19.00	20.45	1.3620	1.0756	1.764
20.00	21.62	1.3636	1.0800	1.839
21.00	22.79	1.3653	1.0845	1.920
22.00	23.98	1.3670	1.0889	2.006
23.00	25.17	1.3687	1.0935	2.099
24.00	26.38	1.3704	1.0980	2.200
25.00	27.59	1.3721	1.1026	2.308
26.00	28.81	1.3738	1.1072	2.424
27.00	30.05	1.3756	1.1119	2.550
28.00	31.29	1.3773	1.1166	2.686
29.00	32.54	1.3791	1.1213	2.834
30.00	33.81	1.3809	1.1261	2.994
31.00	35.08	1.3827	1.1309	3.168
32.00	36.37	1.3845	1.1358	3.357
33.00	37.67	1.3863	1.1406	3.564
34.00	38.98	1.3881	1.1455	3.791
35.00	40.29	1.3900	1.1505	4.039

Table B-6 *(Continued)*

Percent w/w	Percent w/v	Refractive Index	Density (gm/ml)	Viscosity (cP)
36.00	41.62	1.3918	1.1555	4.312
37.00	42.97	1.3937	1.1605	4.613
38.00	44.32	1.3956	1.1656	4.945
39.00	45.68	1.3975	1.1706	5.313
40.00	47.06	1.3995	1.1758	5.722
41.00	48.45	1.4014	1.1809	6.178
42.00	49.84	1.4034	1.1861	6.689
43.00	51.26	1.4054	1.1914	7.262
44.00	52.68	1.4074	1.1966	7.907
45.00	54.11	1.4093	1.2019	8.636
46.00	55.56	1.4114	1.2073	9.462
47.00	57.02	1.4134	1.2126	10.40
48.00	58.90	1.4155	1.2180	11.47
49.00	59.98	1.4176	1.2234	12.70
50.00	61.48	1.4197	1.2288	14.11
51.00	62.99	1.4218	1.2343	15.73
52.00	64.51	1.4239	1.2399	17.60
53.00	66.05	1.4261	1.2455	19.80
54.00	67.60	1.4282	1.2512	22.36
55.00	69.16	1.4304	1.2568	25.38
56.00	70.74	1.4326	1.2625	28.96
57.00	72.33	1.4348	1.2682	33.22
58.00	73.94	1.4370	1.2739	38.34
59.00	75.56	1.4393	1.2798	44.53
60.00	77.19	1.4415	1.2856	52.08
61.00	78.83	1.4438	1.2915	61.35
62.00	80.49	1.4461	1.2974	72.83
63.00	82.17	1.4483	1.3033	87.19
64.00	83.86	1.4506	1.3094	105.2
65.00	85.56	1.4529	1.3153	128.3
66.00	87.28	1.4552	1.3215	158.0

Source: National Bureau of Standards (USA) and *Principles of Refractometry,* Bausch and Lomb Analytical Systems Division, Rochester, New York.

Table B-7 Physical Properties of Sucrose in Saline at 4°C

Concentration		Refractive	Density	Viscosity
Percent w/w	Percent w/v	Index	(gm/ml)	(cP)
0	0	1.3361	1.0057	1.58
0.99	1	1.3375	1.0095	1.63
1.97	2	1.3389	1.0133	1.67
2.95	3	1.3403	1.0171	1.72
3.92	4	1.3418	1.0209	1.77
4.88	5	1.3432	1.0247	1.82
5.83	6	1.3446	1.0285	1.87
6.78	7	1.3460	1.0323	1.93
7.72	8	1.3474	1.0362	1.99
8.65	9	1.3488	1.0400	2.06
9.58	10	1.3503	1.0437	2.13
10.50	11	1.3517	1.0476	2.20
11.41	12	1.3531	1.0514	2.28
12.32	13	1.3545	1.0552	2.36
13.22	14	1.3559	1.0590	2.44
14.11	15	1.3573	1.0628	2.52
15.00	16	1.3587	1.0666	2.62
15.88	17	1.3602	1.0704	2.73
16.76	18	1.3616	1.0742	2.84
17.63	19	1.3630	1.0780	2.95
18.49	20	1.3644	1.0818	3.07
19.34	21	1.3658	1.0856	3.19
20.19	22	1.3672	1.0894	3.32
21.04	23	1.3686	1.0932	3.47
21.88	24	1.3701	1.0971	3.62
22.71	25	1.3715	1.1009	3.78
23.54	26	1.3729	1.1046	3.93
24.36	27	1.3743	1.1085	4.13
25.17	28	1.3757	1.1123	4.30
25.98	29	1.3771	1.1161	4.49
26.79	30	1.3786	1.1199	4.69
27.59	31	1.3800	1.1237	4.94

Table B-7   (*Continued*)

| Concentration | | Refractive | Density | Viscosity |
Percent w/w	Percent w/v	Index	(gm/ml)	(cP)
28.38	32	1.3814	1.1275	5.20
29.17	33	1.3828	1.1313	5.47
29.95	34	1.3842	1.1351	5.77
30.73	35	1.3856	1.1389	6.07
31.50	36	1.3870	1.1427	6.39
32.27	37	1.3884	1.1465	6.75
33.03	38	1.3899	1.1503	7.11
33.79	39	1.3913	1.1542	7.51
34.54	40	1.3927	1.1580	7.93
35.29	41	1.3941	1.1618	8.35
36.03	42	1.3955	1.1656	8.85
36.77	43	1.3969	1.1694	9.40
37.50	44	1.3984	1.1732	9.99
38.23	45	1.3998	1.1770	10.67
38.96	46	1.4012	1.1808	11.42
39.68	47	1.4026	1.1846	12.22
40.39	48	1.4040	1.1884	13.05
41.10	49	1.4054	1.1922	13.95
41.81	50	1.4069	1.1960	14.90
42.51	51	1.4083	1.1998	15.90
43.20	52	1.4097	1.2036	17.10
43.89	53	1.4111	1.2074	18.40
44.58	54	1.4125	1.2113	19.80
45.26	55	1.4139	1.2151	21.30
45.94	56	1.4153	1.2189	23.10
46.62	57	1.4167	1.2227	25.10
47.29	58	1.4182	1.2265	27.20
47.96	59	1.4196	1.2303	29.30
48.62	60	1.4210	1.2341	31.40

Source: Milgrom, E., and Sheeler, P. (previously unpublished).

Table B-8 Physical Properties of Sucrose in Saline at 22°C

Concentration		Refractive Index	Density (gm/ml)	Viscosity (cP)
Percent w/w	Percent w/v			
0	0	1.3349	1.0033	0.96
0.99	1	1.3363	1.0070	0.98
1.98	2	1.3378	1.0108	1.00
2.96	3	1.3392	1.0145	1.03
3.93	4	1.3406	1.0183	1.06
4.89	5	1.3421	1.0220	1.09
5.85	6	1.3435	1.0258	1.12
6.80	7	1.3449	1.0295	1.15
7.74	8	1.3463	1.0332	1.19
8.68	9	1.3478	1.0370	1.23
9.61	10	1.3492	1.0407	1.27
10.53	11	1.3506	1.0445	1.31
11.45	12	1.3521	1.0482	1.35
12.36	13	1.3535	1.0519	1.39
13.26	14	1.3549	1.0557	1.43
14.16	15	1.3564	1.0594	1.48
15.05	16	1.3578	1.0632	1.53
15.93	17	1.3592	1.0670	1.58
16.81	18	1.3606	1.0706	1.63
17.68	19	1.3621	1.0744	1.68
18.55	20	1.3635	1.0781	1.73
19.41	21	1.3649	1.0819	1.80
20.27	22	1.3664	1.0856	1.87
21.11	23	1.3678	1.0894	1.95
21.96	24	1.3692	1.0931	2.03
22.79	25	1.3707	1.0968	2.11
23.62	26	1.3721	1.1006	2.19
24.45	27	1.3735	1.1043	2.28
25.27	28	1.3749	1.1081	2.38
26.08	29	1.3764	1.1118	2.49
26.89	30	1.3778	1.1155	2.60
27.70	31	1.3792	1.1193	2.71

Table B-8 (*Continued*)

Concentration		Refractive Index	Density (gm/ml)	Viscosity (cP)
Percent w/w	Percent w/v			
28.50	32	1.3807	1.1230	2.82
29.29	33	1.3821	1.1268	2.93
30.08	34	1.3835	1.1305	3.05
30.86	35	1.3850	1.1343	3.18
31.63	36	1.3864	1.1380	3.32
32.41	37	1.3878	1.1417	3.46
33.17	38	1.3892	1.1455	3.62
33.94	39	1.3907	1.1492	3.80
34.69	40	1.3921	1.1530	4.00
35.45	41	1.3935	1.1567	4.21
36.19	42	1.3950	1.1605	4.44
36.94	43	1.3964	1.1642	4.69
37.68	44	1.3978	1.1679	4.95
38.40	45	1.3993	1.1718	5.22
39.14	46	1.4007	1.1754	5.51
39.86	47	1.4021	1.1792	5.82
40.58	48	1.4035	1.1829	6.14
41.29	49	1.4050	1.1866	6.49
42.00	50	1.4064	1.1904	6.85
42.71	51	1.4078	1.1941	7.26
43.41	52	1.4093	1.1979	7.68
44.11	53	1.4107	1.2016	8.16
44.80	54	1.4121	1.2053	8.66
45.49	55	1.4136	1.2091	9.19
46.17	56	1.4150	1.2128	9.80
46.85	57	1.4164	1.2166	10.48
47.53	58	1.4178	1.2203	11.26
48.20	59	1.4193	1.2241	12.10
48.87	60	1.4207	1.2278	12.99

Source: Milgrom, E., and Sheeler, P. (previously unpublished).

Table B-9 Physical Properties of Aqueous Metrizamide
Solutions at 20°C

Concentration		Refractive	Density	Viscosity
Percent w/w	Percent w/v	Index	(gm/ml)	(cP)
0.00	0.0	1.3330	0.9988	1.00
4.87	5.0	1.3407	1.0250	1.15
9.51	10.0	1.3483	1.0512	1.30
13.90	15.0	1.3564	1.0787	1.45
18.07	20.0	1.3646	1.1062	1.60
22.05	25.0	1.3728	1.1337	1.90
25.83	30.0	1.3809	1.1612	2.30
29.44	35.0	1.3890	1.1887	2.95
32.89	40.0	1.3971	1.2162	3.60
36.18	45.0	1.4052	1.2437	4.80
39.33	50.0	1.4133	1.2712	6.00
42.35	55.0	1.4214	1.2986	8.00
45.24	60.0	1.4295	1.3262	11.0
48.02	65.0	1.4376	1.3536	16.0
50.68	70.0	1.4458	1.3812	26.0
53.24	75.0	1.4539	1.4087	42.0
55.70	80.0	1.4620	1.4362	58.0

Source: Rickwood, D. (1977), *Metrizamide, A Gradient Medium for Centrifugation Studies.* Nyegaard, Oslo, Norway.

Table B-10 Physical Properties of Metrizamide in Saline at 4°C

Concentration		Refractive Index	Density (gm/ml)	Viscosity (cP)
Percent w/w	Percent w/v			
0	0	1.3361	1.0057	1.58
0.99	1	1.3376	1.0109	1.61
1.97	2	1.3390	1.0160	1.64
2.94	3	1.3405	1.0212	1.67
3.90	4	1.3420	1.0263	1.71
4.85	5	1.3435	1.0316	1.75
5.79	6	1.3449	1.0367	1.79
6.72	7	1.3464	1.0419	1.84
7.64	8	1.3479	1.0471	1.89
8.55	9	1.3494	1.0522	1.94
9.46	10	1.3508	1.0574	1.99
10.35	11	1.3523	1.0626	2.05
11.24	12	1.3538	1.0678	2.11
12.12	13	1.3552	1.0729	2.17
12.99	14	1.3567	1.0781	2.23
13.85	15	1.3582	1.0833	2.30
14.70	16	1.3597	1.0884	2.37
15.47	17	1.3611	1.0936	2.44
16.38	18	1.3626	1.0988	2.52
17.21	19	1.3641	1.1040	2.60
18.03	20	1.3656	1.1091	2.68
18.85	21	1.3670	1.1143	2.77
19.65	22	1.3685	1.1195	2.87
20.45	23	1.3700	1.1246	2.97
21.24	24	1.3715	1.1298	3.09
22.03	25	1.3729	1.1350	3.21
22.70	26	1.3744	1.1402	3.33
23.57	27	1.3759	1.1453	3.46
24.34	28	1.3773	1.1505	3.59
25.09	29	1.3788	1.1557	3.72
25.84	30	1.3803	1.1609	3.85
26.59	31	1.3818	1.1660	3.99

Table B-10 *(Continued)*

| Concentration | | Refractive | Density | Viscosity |
Percent w/w	Percent w/v	Index	(gm/ml)	(cP)
27.32	32	1.3832	1.1712	4.15
28.05	33	1.3847	1.1764	4.32
28.78	34	1.3862	1.1815	4.49
29.49	35	1.3877	1.1867	4.68
30.20	36	1.3891	1.1919	4.91
30.91	37	1.3906	1.1971	5.14
31.61	38	1.3921	1.2022	5.38
32.30	39	1.3936	1.2074	5.64
32.99	40	1.3950	1.2126	5.92
33.67	41	1.3965	1.2177	6.21
34.34	42	1.3980	1.2229	6.53
35.01	43	1.3994	1.2281	6.86
35.68	44	1.4009	1.2333	7.23
36.34	45	1.4024	1.2384	7.61
36.99	46	1.4039	1.2436	8.00
37.64	47	1.4053	1.2488	8.45
38.28	48	1.4068	1.2539	8.93
38.92	49	1.4083	1.2591	9.45
39.55	50	1.4098	1.2643	9.99
40.17	51	1.4112	1.2695	10.55
40.80	52	1.4127	1.2746	11.13
41.42	53	1.4142	1.2797	11.73
42.02	54	1.4157	1.2850	12.35
42.63	55	1.4171	1.2901	12.99
43.23	56	1.4186	1.2953	13.66
43.83	57	1.4201	1.3005	14.35
44.42	58	1.4216	1.3057	15.06
45.01	59	1.4230	1.3108	15.86
45.59	60	1.4245	1.3160	16.80

Source: Milgrom, E., and Sheeler, P. (previously unpublished).

Table B-11 Physical Properties of Metrizamide in Saline at 22°C

Percent w/w	Percent w/v	Refractive Index	Density (gm/ml)	Viscosity (cP)
0	0	1.3349	1.0033	0.96
0.99	1	1.3364	1.0084	0.98
1.97	2	1.3379	1.0135	1.00
2.95	3	1.3393	1.0185	1.02
3.91	4	1.3408	1.0236	1.04
4.86	5	1.3423	1.0287	1.06
5.80	6	1.3438	1.0338	1.08
6.74	7	1.3453	1.0388	1.10
7.66	8	1.3468	1.0439	1.13
8.58	9	1.3482	1.0490	1.16
9.49	10	1.3497	1.0541	1.19
10.39	11	1.3512	1.0592	1.22
11.28	12	1.3527	1.0642	1.25
12.16	13	1.3542	1.0693	1.28
13.03	14	1.3556	1.0744	1.31
13.90	15	1.3571	1.0795	1.34
14.75	16	1.3586	1.0846	1.37
15.60	17	1.3601	1.0896	1.41
16.44	18	1.3616	1.0947	1.45
17.28	19	1.3631	1.0998	1.49
18.10	20	1.3645	1.1049	1.53
18.92	21	1.3660	1.1099	1.57
19.73	22	1.3675	1.1150	1.62
20.53	23	1.3690	1.1206	1.67
21.33	24	1.3705	1.1252	1.72
22.12	25	1.3719	1.1303	1.77
22.90	26	1.3734	1.1353	1.83
23.68	27	1.3749	1.1404	1.89
24.44	28	1.3764	1.1455	1.96
25.20	29	1.3779	1.1506	2.03
25.96	30	1.3793	1.1556	2.10
26.71	31	1.3808	1.1607	2.18

Table B-11   (*Continued*)

Concentration		Refractive	Density	Viscosity
Percent w/w	Percent w/v	Index	(gm/ml)	(cP)
27.45	32	1.3823	1.1658	2.26
28.18	33	1.3838	1.1709	2.34
28.91	34	1.3853	1.1760	2.42
29.64	35	1.3868	1.1810	2.50
30.35	36	1.3882	1.1861	2.58
31.06	37	1.3897	1.1912	2.67
31.77	38	1.3912	1.1963	2.76
32.46	39	1.3927	1.2014	2.86
33.16	40	1.3942	1.2064	2.97
33.84	41	1.3956	1.2115	3.09
34.52	42	1.3971	1.2166	3.21
35.20	43	1.3986	1.2217	3.35
35.87	44	1.4000	1.2267	3.50
36.53	45	1.4016	1.2318	3.67
37.19	46	1.4030	1.2369	3.85
37.84	47	1.4045	1.2420	4.03
38.49	48	1.4060	1.2470	4.23
39.13	49	1.4075	1.2521	4.45
39.77	50	1.4090	1.2572	4.68
40.40	51	1.4105	1.2623	4.92
41.03	52	1.4119	1.2674	5.17
41.65	53	1.4134	1.2725	5.44
42.27	54	1.4149	1.2775	5.72
42.88	55	1.4164	1.2826	6.04
43.49	56	1.4179	1.2877	6.40
44.09	57	1.4194	1.2928	6.78
44.69	58	1.4208	1.2978	7.18
45.28	59	1.4223	1.3029	7.60
45.87	60	1.4238	1.3080	8.10

Source: Milgrom, E., and Sheeler, P. (previously unpublished).

Table B-12 Physical Properties of Aqueous Ficoll Solutions at 4°C

Concentration		Refractive Index	Density (gm/ml)	Viscosity (cP)
Percent w/w	Percent w/v			
0.00	0.000	1.3346	1.0004	1.564
3.00	3.032	1.3382	1.0106	2.788
4.00	4.058	1.3392	1.0145	3.397
5.00	5.090	1.3408	1.0180	4.102
6.00	6.129	1.3420	1.0215	5.014
7.00	7.177	1.3437	1.0253	6.017
8.00	8.232	1.3454	1.0290	7.372
9.00	9.295	1.3469	1.0328	8.579
10.00	10.37	1.3484	1.0365	10.35
12.00	12.53	1.3514	1.0441	14.27
14.00	14.73	1.3538	1.0518	20.21
16.00	16.96	1.3579	1.0597	27.42
18.00	19.21	1.3608	1.0673	38.33
20.00	21.50	1.3645	1.0752	52.31
22.00	23.84	1.3680	1.0837	69.28
24.00	26.21	1.3714	1.0922	95.03
26.00	28.61	1.3748	1.1004	125.7
28.00	31.06	1.3786	1.1093	172.0
30.00	33.53	1.3820	1.1176	225.6
32.00	36.04	1.3856	1.1263	308.8
34.00	38.57	1.3890	1.1345	407.2
36.00	41.19	1.3930	1.1442	565.4
38.00	43.83	1.3970	1.1534	762.2
40.00	46.52	1.4030	1.1629	1020.0

Source: Pretlow, T. G., Boone, C. W., Shrager, R. I., and Weiss, G. H. (1969) Rate zonal centrifugation in a Ficoll gradient. *Anal. Biochem.* **29,** 230.

Table B-13   Physical Properties of Aqueous Ficoll Solutions at 20°C

Concentration		Refractive Index	Density (gm/ml)	Viscosity (cP)
Percent w/w	Percent w/v			
0.00	0.0	1.3330	0.999	1.0
4.92	5.0	1.3402	1.016	2.0
9.67	10.0	1.3474	1.034	5.0
14.27	15.0	1.3547	1.051	7.0
18.74	20.0	1.3619	1.067	20.0
23.06	25.0	1.3692	1.084	35.0
27.27	30.0	1.3764	1.102	60.0
31.33	35.0	1.3836	1.119	104
35.33	40.0	1.3909	1.136	180
39.19	45.0	1.3981	1.153	340
43.10	50.0	1.4053	1.170	600

Source: *Ficoll,* Pharmacia Fine Chemicals AB, Box 175, S-751 04, Uppsala 1, Sweden.

Table B-14   Physical Properties of Ficoll in Saline at 4°C

| Concentration | | | | |
Percent w/w	Percent w/v	Refractive Index	Density (gm/ml)	Viscosity (cP)
0	0	1.3361	1.0057	1.58
0.99	1	1.3375	1.0090	1.91
1.98	2	1.3389	1.0124	2.27
2.95	3	1.3402	1.0158	2.74
3.93	4	1.3417	1.0192	3.23
4.89	5	1.3431	1.0226	3.79
5.85	6	1.3445	1.0259	4.44
6.80	7	1.3459	1.0293	5.21
7.75	8	1.3473	1.0327	6.16
8.69	9	1.3487	1.0360	7.41
9.62	10	1.3501	1.0394	8.84
10.55	11	1.3515	1.0428	10.48
11.47	12	1.3529	1.0462	12.30
12.39	13	1.3542	1.0495	14.30
13.30	14	1.3556	1.0529	16.80
14.20	15	1.3570	1.0563	19.85
15.10	16	1.3584	1.0597	23.00
15.99	17	1.3598	1.0630	27.10
16.88	18	1.3612	1.0664	31.40
17.76	19	1.3626	1.0698	35.90
18.64	20	1.3640	1.0731	40.70
19.51	21	1.3654	1.0765	45.90
20.37	22	1.3668	1.0799	51.50
21.23	23	1.3682	1.0833	57.50
22.09	24	1.3696	1.0866	64.60
22.94	25	1.3710	1.0900	72.50

Source: Milgrom, E., and Sheeler, P. (previously unpublished).

Table B-15 Physical Properties of Ficoll in Saline at 22°C

Concentration		Refractive Index	Density (gm/ml)	Viscosity (cP)
Percent w/w	Percent w/v			
0	0	1.3349	1.0033	0.96
0.99	1	1.3363	1.0066	1.12
1.98	2	1.3377	1.0099	1.31
2.96	3	1.3392	1.0132	1.53
3.94	4	1.3406	1.0165	1.78
4.90	5	1.3420	1.0199	2.06
5.86	6	1.3434	1.0232	2.38
6.82	7	1.3448	1.0265	2.78
7.77	8	1.3463	1.0298	3.30
8.71	9	1.3477	1.0331	3.95
9.65	10	1.3491	1.0364	4.80
10.58	11	1.3505	1.0397	5.80
11.51	12	1.3519	1.0430	7.00
12.42	13	1.3534	1.0464	8.30
13.34	14	1.3548	1.0497	9.70
14.25	15	1.3562	1.0530	11.20
15.15	16	1.3576	1.0563	12.80
16.04	17	1.3590	1.0596	14.50
16.94	18	1.3605	1.0629	16.30
17.82	19	1.3619	1.0662	18.20
18.70	20	1.3633	1.0695	20.20
19.57	21	1.3647	1.0729	22.40
20.44	22	1.3661	1.0762	24.90
21.31	23	1.3676	1.0795	27.90
22.17	24	1.3690	1.0828	31.90
23.02	25	1.3704	1.0861	36.90

Source: Milgrom, E., and Sheeler, P. (previously unpublished).

Table B-16   Physical Properties of Aqueous Percoll Solutions at
20°C

Concentration				
Percent w/w	Percent w/v	Refractive Index	Density (gm/ml)	Viscosity (cP)
0.0	0.00	1.3330	0.999	1.00
1.0	1.01	1.3339	1.005	1.17
2.0	2.02	1.3348	1.011	1.27
3.0	3.05	1.3357	1.016	1.38
4.0	4.09	1.3367	1.022	1.50
5.0	5.14	1.3376	1.028	1.62
6.0	6.20	1.3386	1.034	1.75
7.0	7.27	1.3395	1.039	1.92
8.0	8.36	1.3404	1.045	2.10
9.0	9.46	1.3413	1.051	2.29
10.0	10.57	1.3422	1.057	2.54
11.0	11.68	1.3432	1.062	2.80
12.0	12.82	1.3441	1.068	3.13
13.0	13.96	1.3450	1.074	3.52
14.0	15.12	1.3460	1.080	3.95
15.0	16.28	1.3469	1.085	4.45
16.0	17.46	1.3478	1.091	5.00
17.0	18.65	1.3487	1.097	5.60
18.0	19.85	1.3497	1.103	6.25
19.0	21.07	1.3506	1.109	6.96
20.0	22.28	1.3515	1.114	7.75
21.0	23.52	1.3524	1.120	8.60
22.0	24.77	1.3534	1.126	9.40
22.7	25.65	1.3540	1.130	10.0

Source: *Percoll for Density Gradient Centrifugation,* Pharmacia
Fine Chemicals, AB. Box 175, S-751 04, Uppsala 1, Sweden; Pertoft, H., Laurent, T. C., Laas, T., and Kagedal, L. (1978) Density
gradients prepared from colloidal silica particles coated by polyvinylpyrrolidone (Percoll). *Anal. Biochem.* **88,** 21.

Table B-17   Physical Properties of Percoll in Saline at 4°C

Concentration				
Percent w/w	Percent w/v	Refractive Index	Density (gm/ml)	Viscosity (cP)
0	0	1.3361	1.0057	1.58
0.99	1	1.3368	1.0109	1.62
1.97	2	1.3375	1.0162	1.66
2.94	3	1.3383	1.0214	1.70
3.90	4	1.3390	1.0266	1.74
4.85	5	1.3397	1.0319	1.79
5.79	6	1.3404	1.0371	1.84
6.72	7	1.3412	1.0423	1.89
7.64	8	1.3419	1.0476	1.94
8.55	9	1.3426	1.0528	2.00
9.45	10	1.3433	1.0580	2.06
10.35	11	1.3441	1.0632	2.12
11.23	12	1.3448	1.0685	2.18
12.11	13	1.3455	1.0737	2.25
12.98	14	1.3462	1.0789	2.32
13.84	15	1.3470	1.0842	2.40
14.69	16	1.3476	1.0894	2.48
15.53	17	1.3484	1.0946	2.57
16.37	18	1.3491	1.0999	2.66
17.19	19	1.3499	1.1051	2.76
18.01	20	1.3506	1.1103	2.86
18.82	21	1.3513	1.1156	2.97
19.63	22	1.3520	1.1208	3.09
20.43	23	1.3528	1.1260	3.21
21.21	24	1.3535	1.1313	3.33
22.00	25	1.3542	1.1365	3.46

Source: Milgrom, E., and Sheeler, P. (previously unpublished).

Table B-18 Physical Properties of Percoll in Saline at 22°C

Concentration		Refractive	Density	Viscosity
Percent w/w	Percent w/v	Index	(gm/ml)	(cP)
0	0	1.3349	1.0033	0.96
0.99	1	1.3356	1.0085	0.97
1.97	2	1.3364	1.0136	0.98
2.95	3	1.3371	1.0188	0.99
3.91	4	1.3379	1.0240	1.01
4.86	5	1.3386	1.0291	1.03
5.80	6	1.3393	1.0343	1.05
6.73	7	1.3401	1.0395	1.07
7.66	8	1.3408	1.0446	1.09
8.57	9	1.3416	1.0498	1.12
9.48	10	1.3423	1.0550	1.15
10.38	11	1.3430	1.0601	1.18
11.26	12	1.3438	1.0653	1.22
12.14	13	1.3445	1.0705	1.26
13.02	14	1.3453	1.0757	1.30
13.88	15	1.3460	1.0808	1.34
14.73	16	1.3467	1.0860	1.38
15.58	17	1.3475	1.0912	1.43
16.42	18	1.3482	1.0963	1.48
17.25	19	1.3490	1.1015	1.53
18.07	20	1.3497	1.1067	1.58
18.89	21	1.3504	1.1118	1.64
19.70	22	1.3512	1.1170	1.70
20.50	23	1.3519	1.1222	1.77
21.29	24	1.3527	1.1273	1.85
22.08	25	1.3534	1.1325	1.94

Source: Milgrom, E., and Sheeler, P. (previously unpublished).

List of Manufacturers of Centrifuges, Rotors, and Centrifugation Accessories

Company (Alphabetical Order)	Equipment–materials
Bausch and Lomb 820 Linden Avenue Rochester, New York 14625	Refractometers
Beckman Instruments, Inc. Spinco Division 1117 California Avenue Palo Alto, California 94304	Centrifuges Rotors Gradient makers Gradient materials
Buchler Instruments Division of Searle Diagnostics, Inc. 1327 Sixteenth Street Fort Lee, New Jersey 07024	Gradient makers Pumps
Damon/IEC 300 Second Avenue Needham Heights, Massachusetts 02194	Centrifuges Rotors
DuPont/Sorvall Instruments DuPont Company, Instrument Products Division Peck's Lane Newtown, Connecticut 06470	Centrifuges Rotors Gradient makers Pumps
MSE Scientific Instruments Ltd. Manor Royal Crawley, Sussex, United Kingdom	Centrifuges Rotors Gradient makers

264

Nyegaard and Company, AS
P. O. Box 4220
Oslo 4, Norway
(USA Distributor, Accurate Chemical
  and Scientific Corp.)
28 Tec Street
Hicksville, New York 11801

Gradient materials (e.g., metrizamide)

Payton Associates, Inc.
244 Delaware Avenue
Buffalo, New York 14202

Sta-Put apparatus

Pennwalt Corporation
Sharples-Stokes Division
1415 Rollins Road
Burlingame, California 94010

Centrifuges
Rotors

Pharmacia Fine Chemicals AB
P. O. Box 175 S-751 64
Uppsala 1, Sweden
(USA Distributor, Pharmacia Fine
  Chemicals)
800 Centennial Avenue
Piscataway, New Jersey 08854

Gradient materials (e.g., Ficoll, Percoll, Dextran)

Index

267